edexcel
advancing learning, changing lives

Construction

BTEC First Core Units

Sue Meredith

Contributing authors:
Andrew Williams, John Blaus
and Rex Witts

Published by
Edexcel Limited
One90 High Holborn
London
WC1V 7BH

www.edexcel.org.uk

Distributed by
Pearson Education Limited
Edinburgh Gate
Harlow
Essex
CM20 2JE

www.longman.co.uk

© **Edexcel Limited 2007**

First published 2007
Second impression 2008
British Library Cataloguing in Publication Data is
available from the British Library on request

ISBN 978-1-84690-179-9

Commissioned by Jenni Johns
Design and publishing services by Steve Moulds of
DSM Partnership
Cover image courtesy of (top) Construction
photography.com, (bottom) JupiterImages
Project managed by Julia Bruce
Picture research by Thelma Gilbert
Index by Joan Dearnley
Printed and bound by Graficas Estella, Bilboa, Spain

This material offers high quality support for the
delivery of Edexcel qualifications. This does not
mean that it is essential to use it to achieve any
Edexcel qualification, nor does it mean that this is
the only suitable material available to support any
Edexcel qualification. No Edexcel-published material
will be used verbatim in setting any Edexcel
assessment and any resource lists produced by
Edexcel shall include this and other appropriate
texts.

Acknowledgements
Alamy pp 8 (Paul Glendell); 9 (Photofusion); 12
(Shaun Higson); 13 (Ashley Cooper); 24 (Michael
Juno); 31 (Henry Westheim); 33(b) (ImageState); 35
(David R Frazier Photolibrary); 37 (Andrew Fox); 38
(Photofusion); 44(b) (Photofusion); 49 Jinny
Goodman); 50 (David J. Green); 54 (Chris
Spradbery); 57 (Woodystock); 63 (Russ Widstrand);
64 (Michael Harder); 85 (David Lyons).
Constructionphotography.com pp 17; 18; 19; 32;
33(t); 40; 44(t); 53; 56; 65; 83. **Corbis** pp 20; 59.
Ecopaints p9. **Genesis Centre** p45. **Neath Port
Talbot County Borough Council** p15. **Rex Features**
pp 43; 52. **Roger Scruton** pp 22; 99; 135. **Science
Photo Library** p67. **Screwfix** (www.screwfix.com)
p68. **Skyscan**.co.uk p7.

Contents

Welcome to *Construction BTEC First Core Units*. This book will provide you with all the information you need to complete the core units 1, 2 and 3 of the qualification.

This book is organised by topic. Each topic will run for 2, 4, 6 or 8 pages. Essential information that you need to take away with you once you have completed the topic is featured in the 'Need to know' column on the last page.

❋ 'Go to' will refer you to further information on the web or the accompanying CD.

❋ 'Huh?' will define terms used in the topic.

❋ 'Top Tips' will give you some useful pointers to remember.

❋ 'Checked' is a checklist of the main things you need to know.

Other features in the text will also help your learning and understanding.

❋ 'Low down' gives you in-depth explanations of some of the more important themes and topics.

❋ 'On file' case studies give you real-world examples of construction projects, problems and issues.

❋ 'Fact is' summarises important aspects of the topic.

❋ 'Brick talk' gives practical encouragement and advice.

❋ 'Ask' suggests some practical things that you can do to find out more.

NEED TO KNOW...

For information on the web about **diversity in the construction industry** and **organisational structures** see the resources on the CD accompanying this book.

HUH?

Brownfield site – an urban site that has been built on before

Tender – offer a job for people/companies to bid for

Delegate – pass the responsibility for a task or job to another person or organisation

TOP TIPS

✳ If your organisation includes workers in decision-making – get involved

✳ Ask your colleagues for help and advice

CHECKED

ONE	An introduction to the construction industry	BTEC FIRST
SUBJECT	Construction project teams and teamwork	

➲ Construction team members must: understand the goals and timelines of the project, work efficiently and to a high standard, communicate honestly and openly, especially about errors and problems, and trust and respect the client and other team members

➲ Communication about project progress should be regular, frequent, clear and simple

DIFFICULTY RATING	DATE
⛑⛑⛑⛑⛑	

CHECK IT OUT

Find out what sustainable building products are available locally, for example items made from recycled materials.

Find out which local building projects are using sustainable building principles and how these are applied.

Remember: the solutions to any problems that arise must be suitable for the whole project, not just one section.

BRICK TALK

A S K **Conduct a short survey of construction industry professionals, asking the following question:**

'In terms of sustainability and the environment, what is the duty of the construction industry to the present community and to future generations?'

In Unit 3 you will find some special features to help you with the maths and science that underpins construction work.

❋ 'Concepts' describes the basic rules and formulae

❋ 'Ground rules' takes you step-by-step through calculations giving worked examples and a chance to try them out for yourself.

❋ 'Watch it' will help you avoid common pitfalls.

❋ 'In the real world' shows you how the maths and science relates to real-life problems and situations in the construction industry.

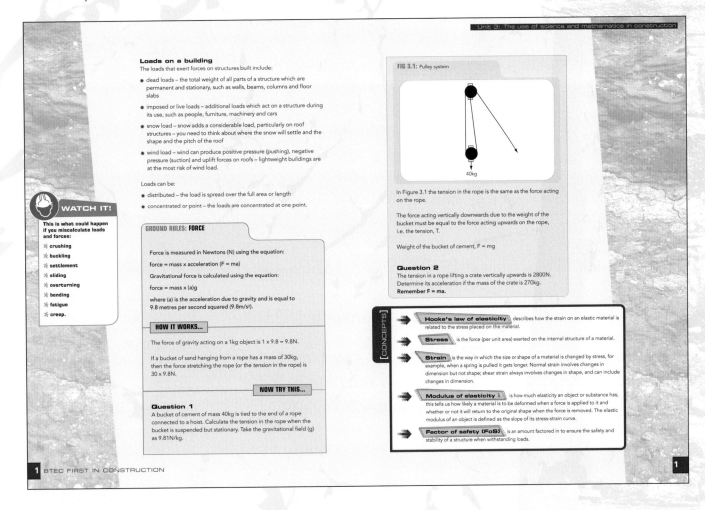

Loads on a building
The loads that exert forces on structures built include:

❋ dead loads – the total weight of all parts of a structure which are permanent and stationary, such as walls, beams, columns and floor slabs

❋ imposed or live loads – additional loads which act on a structure during its use, such as people, furniture, machinery and cars

❋ snow load – snow adds a considerable load, particularly on roof structures – you need to think about where the snow will settle and the shape and the pitch of the roof

❋ wind load – wind can produce positive pressure (pushing), negative pressure (suction) and uplift forces on roofs – lightweight buildings are at the most risk of wind load.

Loads can be:

❋ distributed – the load is spread over the full area or length

❋ concentrated or point – the loads are concentrated at one point.

WATCH IT!

This is what could happen if you miscalculate loads and forces:

❋ crushing
❋ buckling
❋ settlement
❋ sliding
❋ overturning
❋ bending
❋ fatigue
❋ creep.

GROUND RULES: FORCE

Force is measured in Newtons (N) using the equation:

force = mass x acceleration (F = ma)

Gravitational force is calculated using the equation:

force = mass x (a)g

where (a) is the acceleration due to gravity and is equal to 9.8 metres per second squared (9.8m/s²).

HOW IT WORKS...

The force of gravity acting on a 1kg object is 1 x 9.8 = 9.8N.

If a bucket of sand hanging from a rope has a mass of 30kg, then the force stretching the rope (or the tension in the rope) is 30 x 9.8N.

NOW TRY THIS...

Question 1
A bucket of cement of mass 40kg is tied to the end of a rope connected to a hoist. Calculate the tension in the rope when the bucket is suspended but stationary. Take the gravitational field (g) as 9.81N/kg.

FIG 3.1: Pulley system

40kg

In Figure 3.1 the tension in the rope is the same as the force acting on the rope.

The force acting vertically downwards due to the weight of the bucket must be equal to the force acting upwards on the rope, i.e. the tension, T.

Weight of the bucket of cement, F = mg

Question 2
The tension in a rope lifting a crate vertically upwards is 2800N. Determine its acceleration if the mass of the crate is 270kg.
Remember F = ma.

[CONCEPTS]

→ **Hooke's law of elasticity** describes how the strain on an elastic material is related to the stress placed on the material.

→ **Stress** is the force (per unit area) exerted on the internal structure of a material.

→ **Strain** is the way in which the size or shape of a material is changed by stress, for example, when a spring is pulled it gets longer. Normal strain involves changes in dimension but not shape; shear strain always involves changes in shape, and can include changes in dimension.

→ **Modulus of elasticity λ** is how much elasticity an object or substance has; this tells us how likely a material is to be deformed when a force is applied to it and whether or not it will return to the original shape when the force is removed. The elastic modulus of an object is defined as the slope of its stress-strain curve.

→ **Factor of safety (FoS)** is an amount factored in to ensure the safety and stability of a structure when withstanding loads.

1 BTEC FIRST IN CONSTRUCTION

We hope you enjoy using this book in your studies, and we wish you the very best for your course and future career in construction.

When you have completed this unit, you should be able to:

✔ explain the key factors affecting the development of a sustainable built environment, including:
  – the nature of the built environment and the way it can affect the natural environment
  – the importance of sustainable construction to protect the environment
  – how construction contributes to the economy

✔ describe the diversity and complexity of the construction industry, including:
  – the range of job activity areas
  – the range of clients
  – the range of projects

✔ list the human resources available to the construction industry, including:
  – the roles of construction team members and their interaction
  – the variety of career paths, training and education

✔ explain the factors that influence construction projects, including:
  – the influencing factors at each construction stage
  – respecting the natural environment
  – sustainable construction techniques.

An introduction to the construction industry

Coming up in this unit...

This unit is all about how important the construction industry is to all of us. It provides places to live, work and learn, and **infrastructure** such as roads, power stations, airports and railways.

It is important that buildings provide a healthy and pleasant environment and that they 'fit in' with the natural surroundings. In the construction industry, advances in designs and materials are made almost every day – introducing new technology, materials and techniques. Many of these reduce the negative impact that the construction industry has had on the natural environment in the past.

Construction industry employees work together in teams and aim to finish projects to the best possible standards – from the planners and designers at the beginning of a project, to the builders and engineers who create the structure, and the inspectors who check that it is safe.

The construction industry is critically important to us all

The sustainable built environment

To **sustain** something is to keep it going for a long time. To sustain the natural environment, we must make sure that human actions do not destroy any feature beyond repair. The **built environment** includes every structure created by humans. Structures are essential for human survival and modern life – providing home and shelter; places for work, education, recreation, shopping and health support; roads and rail for travel; and structures for essential services such as water and power supply.

We need to use methods for creating environments for humans which do not:

* negatively affect the cleanliness of air and water

* reduce the area of wilderness and other natural environments required for wildlife habitats and a healthy planet

* over-use resources, such as timber.

Sustainability means making sure that whatever we do to the environment now does not damage it beyond repair for the future. A sustainable built environment is one that does not reduce the viability of the natural environment.

We must protect the natural environment

[LOW DOWN]

→ **The built environment** This term covers anything constructed by humans to support human activity. It includes houses, schools, hospitals, factories, water treatment plants, airports, railway stations and other constructions such as roads, bridges, power stations and power distribution lines, railways and landscape architecture – that is, parks and open spaces designed and constructed by humans.

→ **The natural environment** The natural environment includes open spaces with little historical human interference, such as national parks. Wilderness is the most natural environment – one that has not been altered significantly by human activity.

The move towards more sustainable built environments involves very thoughtful and careful planning, design and implementation of every aspect of the construction process. This includes:

* researching the most suitable location for each construction

* organising an **environmental impact assessment** on the selected locations and making sure its recommendations are carried out

* consulting the **community** in the selected locations and keeping the community informed throughout the planning, design and construction processes

* applying sustainable design principles to the size, shape and layout of the building

* considering the physical, mental and emotional health needs of the users, for example security, recreation, accessibility, affordability and social interaction

* consulting the prospective users of the structure throughout the planning, design and construction processes

* using sustainable materials, products, services, technologies and construction methods, including waste minimisation.

PROTECT AND BEAUTIFY WITH PAINT

Rediscover the joys
of real paints

AURO
NATURAL PAINTS

CHECK IT OUT

Find out what sustainable building products are available locally, for example items made from recycled materials.

Find out which local building projects are using sustainable building principles and how these are applied.

Sustainable products and technologies are being used more and more

Conduct a short survey of construction industry professionals, asking the following question:

'In terms of sustainability and the environment, what is the duty of the construction industry to the present community and to future generations?'

The aim of the construction industry must be to safeguard, maintain, improve and expand the built environment with the minimum impact on the natural environment. In addition to ensuring the sustainability of new construction projects, we also have a duty to repair some of the damage done.

Some methods used in the construction industry to minimise environmental impact include:

* Arup – Sustainable Project Appraisal Routine (SPeAR®)

* Building Research Establishment (BRE) – Environmental Assessment Method (BREEAM).

The Sustainable Project Appraisal Routine (SPeAR®) is a software system developed by Arup and used to assess or show how sustainable a project is. It can be used as part of the design process or to assess a product, process or the whole project. The software covers four key aspects of sustainability:

* environmental protection

* social equity (fairness)

* economic viability (whether it is affordable and worth the money)

* efficient use of natural resources.

BREEAM assessment methods help the construction industry to understand and reduce the environmental impact of construction projects. There are different methods for each stage of the process, including:

* BREEAM Developments – used at the master planning stage

* BREEAM Buildings – for assessing individual buildings at all stages

* BREEAM LCA (Life Cycle Analysis) – looks at the environmental impact of construction materials

* BREEAM Envest – helps assess environmental issues at the design stage

* BREEAM Smartwaste – used during construction.

Significant features of the built environment that sustainable strategies could be applied to include the following:

* **Location**: suitability for, and effect on, landscape and environment or architectural period of surrounding buildings; proximity to amenities and transport **infrastructure**.

* **Height and shape**: structural integrity, **aesthetic** effect on surrounds, effect on the aspect of other buildings (such as blocking views), effect of shadow on surrounds.

* **Access**: for buildings – ramps, lifts, stairs; for roads – wildlife pathways underneath.

* **Open space**: air, light, aesthetics, recreation area, human need for personal space.

* **Energy requirements**: effective building design – use of passive solar heat and light, use of solar or renewable energy sources, insulation features, quality of distribution systems.

* **Suitability for purpose**: effective design and location – meeting the needs of users, long-lasting and low maintenance.

ON FILE

SUSTAINABLE CONSTRUCTION

The Construction Industry Council (CIC) is a membership organisation representing over 500,000 professionals in the construction industry and over 25,000 construction firms. The CIC represents the opinions of a wide range of sectors within the construction industry on a variety of important issues.

The following is an extract from the mission statement of the CIC's Sustainable Development Committee.

'The Committee will work with Government, the Industry and its clients, CIC members, and the research community. The Committee will:
- promote the value of sustainable development in the UK construction industry to meet economic, social, safety and environmental challenges
- provide leadership for construction professionals on sustainable development
- provide a forum for debate and use its influence to **facilitate** change
- influence the development of an industry-wide strategy on sustainable development
- encourage improved awareness of sustainability in the education of all construction professionals and their clients.'

TASK

1. Find out what the CIC means by 'economic, social, safety and environmental challenges'.
2. What does this mission statement tell you about the importance of sustainability to the construction industry? Give reasons for your answer.
3. Explore the CIC Sustainable Development Committee website at www.cic.org.uk/activities/sustainComm.shtml, and prepare a short report on at least three news items from the website.

NEED TO KNOW...

GO TO...

For information on the web about **Arup SPEaR**, **BREEAM** and **sustainable development** see the resources on the CD accompanying this book.

HUH?

Sustain – to keep going for a long time

Aesthetic – pleasing to the eye

Facilitate – make easy, assist

Infrastructure – network of services such as roads, power lines, water pipes

Viability – ability to live and thrive

Environmental impact assessment – survey of the natural environment and how a construction project will affect it

Community – all the people living and working in an area

CHECKED

CHAPTER ONE	An introduction to the construction industry	BTEC FIRST
SUBJECT	The sustainable built environment	

➲ A sustainable built environment does not damage the natural environment

➲ Sustainable construction methods involve careful planning and consultation

➲ Sustainable construction methods include waste minimisation, no pollution of air and water, and use of renewable resources

DIFFICULTY RATING	DATE

The construction industry is important socially and economically

Social importance

Communities need houses, workplaces, schools, hospitals, cultural and leisure centres, shopping centres and the transport, communications, water, waste and power infrastructures that service these structures. The construction industry makes an enormous contribution to society, providing this physical environment for human activities.

The quality of planning, design and construction can have a huge impact on the quality of life and health of the population. Compare housing during the boom in city growth in the **Industrial Revolution** – few open spaces, closely packed houses, no indoor toilets or bathrooms, no gardens and a short life expectancy for the working classes – with new housing developments, where features of the built environment are planned to have a positive effect on human health and well-being and the least possible impact on the natural environment. The same is true of large and national infrastructure projects, for example road systems and electricity generating projects. The construction industry has come a long way – but there is still quite a way to go. As the population grows and technology advances and environmental concerns become more urgent, careful planning for future construction needs is essential.

Victorian workers' houses were closely packed, with only basic facilities

[LOW DOWN]

Gross Domestic Product (GDP) The GDP measures the activity of the national economy by assessing the amount of goods and services produced, and the amount of income generated and money spent. The contribution of each industry sector to the GDP can also be measured. Recent estimates of the contribution of the construction industry to the GDP are about 10 per cent.

Pollution Types of pollution that may be caused by the construction industry and the built environment include: noise pollution (especially during construction), air pollution (dust, smoke), soil pollution (chemicals), water pollution (waste, chemicals), visual pollution (cranes, waste, power lines) and light pollution (increased illumination at night, shadows cast by buildings).

Modern housing attempts to create a pleasant and environmentally sound living space

Economic importance

The construction industry provides jobs for about three million people, including those:

* working directly in the industry

* extracting materials used by the industry

* manufacturing construction supplies

* providing services used by the industry.

ASK Conduct a small survey, asking members of your group or class, or of your local community, the following questions:

What are the benefits and drawbacks of:

✱ closing small hospitals and building one new large hospital?

✱ closing small local shops and building one new large shopping centre?

THINK?

What do you think the effect of over-development would be on the following:

✱ Parking?

✱ Traffic congestion?

✱ Pollution?

✱ Property prices?

What else could be affected by over-development? Write some notes on your thoughts, or discuss them in your group or class.

Most jobs in construction and construction-related industries are well paid. These wages and salaries are used to buy property, goods and services, creating a **cash flow** into other areas of the **economy**. In this way, the construction industry makes a large contribution to the **Gross Domestic Product** (GDP). Industrial and commercial buildings that are thoughtfully planned, designed and created by the construction industry attract new businesses from other parts of the country, and even overseas, to move in and set up in an area. As more businesses are attracted to an area, more jobs become available, more money is paid to workers from the local community, and the economy thrives. This is good for the local community and the nation as a whole. The profile of the area is raised within the business community and it becomes a 'happening place'.

Another economic factor is the ongoing value of the building. Most buildings increase in value because of the continuing rise in property prices. Buildings that have been well planned, designed and built will increase in value at a higher rate than other buildings. Good construction is a good investment.

Impacts of the built environment

The good and bad impacts of construction can be summarised as follows:

* Upside:
  - construction of homes and workplaces
  - provision of facilities, such as hospitals, schools and for recreation
  - creation of national infrastructure – road, rail, power, water, waste
  - attracts new business
  - job creation
  - contribution to the economy.

* Downside:
  - reduction of the area of land kept as natural environment
  - increase in **pollution** from the construction process, traffic and industry
  - poor design, over-development and large numbers of people living in **urban** areas may increase ill-health and crime.

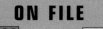

ON FILE

BAGLAN ENERGY PARK

South Wales has lost a lot of employment in pollution-producing industries such as coal-mining, oil refining and manufacturing. For example, the BP oil refinery at Port Talbot closed with a loss of 20,000 jobs. A site was recently chosen for the development of a 180-acre energy park near the M4 at Port Talbot. The whole extent of the Baglan Bay redevelopment site will eventually be 1500 acres – the biggest industrial development site in the UK. The development **consortium** has built a gas-fired power plant for the site, and already companies such as Hi-Lex Cable and Remploy have moved in. The development consortium is hoping that the site will provide 10,000 jobs for the local community and add up to £150 million a year to the local economy.

TASK

Look at the Baglan Energy park website:
www.npt.gov.uk/baglanenergypark/index.cfm
and complete the following activities:

1. Look at the photo gallery of completed buildings at the site. Select the building that you find the most attractive. Draw a sketch of the building and write notes about why you chose this one?

2. Check the 'How to get here' section of the website. What do you think of the location? Is it accessible to all? How many forms of transport can access the site?

3. List the benefits and drawbacks of the closure of the BP oil refinery.

4. List the benefits of the Baglan Energy Park to the local community.

CHECKED

CHAPTER ONE	An introduction to the construction industry	BTEC FIRST
SUBJECT	The construction industry is important socially and economically	

➲ **Social importance: provides buildings and infrastructure for human activities**

➲ **Economic importance: provides jobs in construction and manufacturing; creates buildings that attract businesses to an area**

➲ **Environmental impact: pollution, loss of natural landscape, over-development**

DIFFICULTY RATING	DATE

What's going on in the construction industry

The term 'construction industry' covers a wide variety of construction types, such as residential, **commercial** or industrial buildings, bridges and roads. Within each of these sectors there is a range of activities, such as planning and design, new build, **restoration** and **renovation**, and maintenance and repair. Then, within the activity groups, there are different types of job and skill level. Where do you fit in?

Activity area – Planning: what goes where?

Town and country planning involves working out where buildings, roads, power stations and other structures will be built. The Royal Town Planning Institute (RTPI) promotes good planning in the UK. Good planning should take into consideration **spatial** aspects (space and place), sustainability, **integration** with existing structures and use, and future plans. Town and country planning should also include extensive consultation with all **stakeholders**.

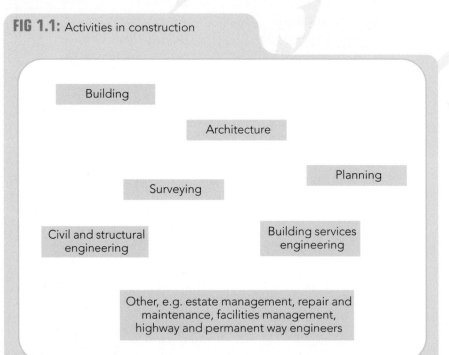

FIG 1.1: Activities in construction

Building

Architecture

Planning

Surveying

Civil and structural engineering

Building services engineering

Other, e.g. estate management, repair and maintenance, facilities management, highway and permanent way engineers

Activity area – Architecture: what will it look like?

Architects are imaginative and creative – designing new buildings and drawing plans for restoring old buildings. Architects keep a close eye on the building part of a project to ensure the design is followed and is working. They also consult with both the client and the contractors, and negotiate any changes. To be a good architect you need an artistic imagination and the ability to draw in three dimensions.

Activity area – Surveying: does it meet legal requirements?

Building surveyors understand the laws, regulations, standards and codes of practice that apply to construction. They supervise projects for organisations such as local councils, conservation bodies or large construction firms – making sure everything is done correctly. Building surveyors also give advice to builders, owners and facilities managers on details for the design, construction, maintenance, repair, renovation and conservation of all types of building.

THINK?

Look at each of the activity areas described in this unit and place them in order, based on how much you would like a job in each area.

Which activity area would you most like to work in? Why?

Which activity area would you least like to work in? Why?

Architect

Activity area – Civil and structural engineering: will it stay up?

Civil engineers plan, manage, design and supervise construction or maintenance projects on fixed structures such as bridges, power plants, roads, railways, dams and flood management structures. Structural engineering is an area of civil engineering concerned with ensuring that constructions can withstand all the weight, stresses and pressures exerted on them. Stresses include natural elements such as wind and **seismic** forces. You'll learn more about stresses in Unit 3.

Bricklayer

Activity area – Building crafts: making the design real

Building craft skills include:

* **trowel occupations** – laying bricks, blocks, lintels, stone and mortar to build walls, chimney stacks and archways

* **plastering** – finishing off walls, both inside and outside, with plaster or mixtures of sand and cement (render) and pebble-dash

* **tiling** – applying ceramic, terracotta, stone, granite and marble tiles on walls and floors

* **carpentry and joinery** – making and installing floorboards, cupboards, doors and window frames, and installing roof trusses and partitions

* **roofing** – slating and tiling, flat roof work with felt and mastic asphalt, roof sheeting and cladding, lead sheeting, thatching

* **painting** – finishing walls, floors, doors and fixtures with paint, varnish or wallpaper.

Activity area – Building services engineering: does it have water, sanitation, lighting, heating and cooling?

Building services engineers make buildings comfortable, energy-efficient and safe. Heating, ventilation, air conditioning, lighting, power, telecommunications, plumbing, drainage, fire protection and even the installation of noise management material are completed by workers in the building services engineering area. Apprenticeships in this activity area include plumbing, electrotechnical services, and heating, ventilating, air conditioning and refrigeration (HVACR) services. Most building services engineers are self-employed in owner-operated businesses.

Activity area – Estate management: what needs to be done?

An estate manager looks after the land and property belonging to a large landowner, such as a local council or health authority, or a wealthy family. The work involves day-to-day management of properties; buying, selling and letting or leasing land or property; and advising on land use and new-build projects. Estate managers should have excellent business skills and a good knowledge of land, environment and construction issues.

Activity area – Facilities management: is it running smoothly?

Making sure everything runs smoothly in a large building, such as an office block or hospital, involves all aspects of the building – inside and out. This includes planning the use of space; organising repairs, maintenance and renovations; hiring and managing contractors for waste removal, cleaning, landscaping and catering; and managing the budgets for each of these activities. This work requires a good technical knowledge of building services and business management skills.

Electrician

ON FILE

BLUEPRINT FOR UK CONSTRUCTION SKILLS 2006

ConstructionSkills forecasts that 87,000 new workers will be recruited into construction jobs each year to fill the skill shortages created by industry growth. There is a very high demand for workers with wood-trade skills and for managers, clerical staff, architects, engineers and other design and technical professionals. Some of the projects fuelling the demand for skilled workers are:

■ Kings Cross redevelopment
■ Shellhaven, Felixstowe and Harwich ports projects
■ East London Line extension
■ Victoria Station redevelopment
■ Thames Gateway
■ public-sector house-building
■ widening of the M1 and M25
■ the five-year national water and sewerage programme
■ nuclear decommissioning
■ Olympic Park and Village.

TASK

1. What is the effect of skills shortages on wages and conditions? Research the salaries offered for at least four different jobs in the construction industry. Find out whether wages have risen over the last few years, and if so by how much.

2. Choose one of the major projects listed above. Research the project and make notes on:
   ■ the proposed start date and duration of the project
   ■ the number of workers likely to be employed
   ■ the areas of activity involved in the project
   ■ the types of job vacancy advertised by the project.

NEED TO KNOW...

GO TO...

For information on the web about **Arup SPEaR**, **BREAM** and **sustainable devleopment** see the resources on the CD accompanying this book.

HUH?

Commercial – relating to business – e.g. shops, warehouses and offices

Integration – combining different parts into a working whole

Restoration – bringing an old building back to its original state, for instance by repairing the brick, stone, timber, plaster and other work

Renovation – restoring an old building to good condition, but not necessarily its original condition often updating materials, appearance or use

Seismic – movement of the earth caused by earthquake tremors or subsidence, or by human action, for instance by explosions

Spatial – to do with space and including dimensions, shape and location

Stakeholders – everyone who has an interest in the project

CHECKED

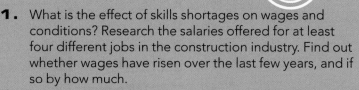

| CHAPTER ONE | An introduction to the construction industry | BTEC FIRST |
| SUBJECT | What's going on in the construction industry | |

➲ Construction industry sectors include: planning and design, new build, restoration and renovation, and maintenance and repair

➲ Construction jobs include: architecture, surveying, civil and structural engineering, building crafts, building services engineering, estate management, facilities management

| DIFFICULTY RATING | DATE |

The wide range of construction projects

As a construction worker you could work on a great variety of projects, ranging from renovating a three-bedroomed house to building a 50-storey apartment block, or from building a warehouse to an entire power station. The range of work carried out by the construction industry covers the following sectors:

* **Residential** – any building which people could call 'home', including:
  – large apartment blocks, such as the Echo 24 project in Sunderland, with 179 luxury apartments
  – housing developments – the demand for new housing in the UK is very high. New developments include brand new estates, both small and large, and conversion of other buildings, such as old schools or chapels, into housing
  – individual residential homes – approximately 17,000 new house-building applications are currently received every month in the UK.

* **Commercial** – buildings used for conducting business, including:
  – large projects, such as the new hangar for the airbus A380, which will use 14,000 tonnes of steel
  – corporate headquarters and regional and local offices
  – small local commercial buildings such as warehouses.

New commercial buildings are often designed to make a huge impact, like this the Oriental Pearl Tower in Shanghai, China

[LOW DOWN]

➡ **Private limited company (Ltd)** A company that does not sell shares to the public. Most small businesses are private companies, usually with one or two directors who own the business and share the profits. Many of these companies start life as family businesses, and family members may own most of the shares. Less legislation and fewer reporting requirements apply to private limited companies.

➡ **Public limited company (plc)** A company that sells shares to the public on the **Stock Exchange**. Most very large companies are public companies. There is usually a board of directors – people who control the company and usually own the largest number of shares. Public companies have to report most of their business dealings and comply with all corporate laws.

✳ **Retail** – really a branch of commercial building, with the use specifically for selling goods and services, including:
 – large shopping centres, for example the Eden Centre in High Wycombe, with 80,000 square metres available for shops, cafés and other retail outlets
 – renovation or **refurbishment** of small high-street shops, including fitting out shop interiors.

✳ **Cultural** – buildings used for displaying art or for artistic performances, including:
 – new theatres and the restoration and maintenance of historic cultural centres such as the Royal Albert Hall
 – art galleries and cultural centres, such as the new cultural centre in Eastbourne, which has galleries, conference and educational facilities, and studios.

✳ **Leisure** – buildings used for sport and entertainment, including:
 – large leisure and sports complexes, such as the Elmbridge Leisure Centre, with swimming pools, sports hall, squash courts, 120-station gym, indoor bowls, climbing wall and many other facilities
 – cinemas.

✳ **Industrial** – buildings used for processing raw materials and manufacturing products, including:
 – large developments such as the Baglan Energy Park
 – new building of small manufacturing and processing facilities
 – refurbishment of old industrial sites.

✳ **Health** – buildings used for medical treatment, including:
 – large hospital complexes such as the new Birmingham Super Hospital, with 1,213 beds and 30 operating theatres
 – residential care homes and nursing homes
 – small local medical centres.

CHECK IT OUT

Which construction sector do the following types of structure belong to?
✳ glasshouse
✳ warehouse
✳ holiday apartments
✳ solicitor's office
✳ dentist
✳ bedsitter
✳ milking shed
✳ yoga studio
✳ multi-storey car park
✳ pub
✳ pharmacist
✳ aqueduct
✳ panel beater's
✳ timber mill
✳ cosmetics manufacturer.

CHECK IT OUT

Consider the structures and buildings in your local area and find out who pays for the construction, renovation or maintenance of: roads, railway station, schools, swimming pool, health centres, rental housing, shops, cinema, county court, town hall.

21

New agricultural buildings should not mar the view

THINK?

Which type of client do you think would be the best to work for? Why?

Which type of project do you think would be the best to work on? Why?

Discuss your opinions with your colleagues and your teacher, trainer or supervisor.

❋ **Educational** – buildings used for teaching and learning, including:
  – 'Building Schools for the Future' – a government project aiming to rebuild or renew every secondary school in England
  – overseas university construction projects such as the University of Liverpool's Xi'an Jiaotong University in China.

❋ **Agricultural** – buildings used for housing animals or storing agricultural products, including:
  – large modern farm buildings, such as metal barns and grain silos – care must be taken to ensure the design suits the landscape, for example not locating buildings on the skyline
  – renovation of traditional farm buildings.

❋ **Utilities and services** – buildings for our electricity, gas and water systems, including:
  – high-voltage substations and electrical power distribution systems
  – liquefied natural gas (LNG) plants, such as the new LNG plant at South Hook, Pembrokeshire
  – water and wastewater treatment plants.

❋ **Public buildings** – buildings used for the administration of the country, including:
  – new office buildings for national and local government
  – restoration of historic buildings, such as the extensive renovation of Birmingham Town Hall.

❋ **Transport infrastructure** – including:
  – roads, railways, bridges and railway stations
  – airports, such as the construction of the new Terminal 5 at Heathrow, which is divided into 16 construction projects, with 147 sub-projects.

ON FILE

THE LARGEST UK CONSTRUCTION COMPANY

Laing O'Rourke is the largest construction company in the UK. It is a private company, run by an **executive** board of four people. The construction division operates in all sectors, with projects in the UK and abroad in the following company divisions:

- lifestyle – including hotels, residential, retail, sport and leisure
- social infrastructure – including defence, education, healthcare, law and order, **urban regeneration**, utilities
- business – including government, industrial, offices, science and research
- transport – including air, marine, rail, roads and bridges.

Some of Laing O'Rourke's recent construction projects include:

- Aintree Racecourse, Liverpool
- Sea Cliff Bridge, New South Wales, Australia
- Channel Tunnel Rail Link, St Pancras Station
- Anglian Water – **potable** and wastewater schemes
- Project Red Dragon – aerospace 'superhangar', Wales
- Snow Centre, Dubai.

In addition to the construction project sectors, the company owns smaller, specialist trading companies offering expertise in: reinforced concrete structures; demolition; piling; precast concrete; pre-stressing and post-tensioning; and stonemasonry.

Laing O'Rourke employs 23,000 people worldwide and offers work experience placements, holiday jobs, apprenticeships, university sponsorships and graduate recruitment. International opportunities are available in Germany, Poland, the United Arab Emirates, India and Australia. Laing O'Rourke is part of the CLM partnership, the winning bidder for the London Olympic Park Development 2012.

TASK

1. Consider the company divisions and types of construction project undertaken by Laing O'Rourke. For each example of the projects listed, suggest the type of client that the company may be working for, such as: hotels – public company.

2. Research the detailed information about construction sectors at Laing O'Rourke listed on the website at:
   http://www.laingorourke.com/sectors_main.htm

   Select two sectors of interest and make notes on the different types of project undertaken by Laing O'Rourke in these sectors. Include details in your notes for each project on:
   - the purpose
   - the size in terms of cost and the number of people employed
   - any specialist skill requirements
   - the timeframe.

NEED TO KNOW...

GO TO...

For information on the web about **Echo 24**, **housebuilding statistics**, **new public and commercial buildings**, **agricultural building** and **Heathrow Terminal 5** see the resources on the CD accompanying this book.

HUH?

Executive – people with an executive role actually run the business

Potable – describes water that is fit for humans to drink

Refurbishment – renewing the internal fittings, fixtures and finishes of a building.

Stock Exchange – organisation based in the City of London where shares in businesses are traded.

Urban regeneration – renewing areas of towns and cities that have become rundown and unattractive

CHECKED

CHAPTER ONE	An introduction to the construction industry	BTEC FIRST
SUBJECT	The wide range of construction projects	

➲ Construction projects include: residential, public, health and educational, commercial and retail, cultural and leisure, industrial and agricultural, utilities and services, transport infrastructure

DIFFICULTY RATING	DATE

Construction project teams and teamwork

Imagine that you own a large **brownfield site** – that is, a site for building that has been used before – near the centre of a city, and you decide to build a sports and leisure complex on it. This is a big construction project, and many different types of skill and expertise will be needed. Who should be involved? Where would you start? How could you get everyone working well together?

Each team member will bring a specific set of skills and expertise to the project, but in addition to this, all construction team members must have:

* a complete and correct understanding of the goals and timelines of the project

* a commitment to the quality standards required and the values of the client, for example to sustainable construction practices

* a commitment to open and honest communication, particularly regarding mistakes, breakdowns and other problems that might affect the project

* trust in, and respect for, the client and other members of the team.

Each of the team members will have particular roles and responsibilities involving constant interaction with other members of the team. It is important that communication about project progress is:

* regular and frequent, for example:
  – a daily update notice on a dedicated website or noticeboard first thing every morning
  – on-site briefings – very short meetings, as required, to discuss and confirm plans

* clear and simple, for example:
  – notices should be in plain English
  – short and to the point.

Brownfield sites are increasingly popular for redevelopment

[LOW DOWN]

Valuing others In an organisation or partnership where all the team members are valued, there is a culture of trust and loyalty, and a strong commitment to making the project successful. Valuing others and being aware of the benefits of diversity in the team involves making a real effort to ensure that everyone is treated equally and fairly. A company that values its workforce should ensure that it provides appropriate and well-maintained equipment, operates in a safe environment, and offers all the training required to do the job well. It should also offer opportunities for personal development and career progression.

Promoting diversity The construction industry continues to work towards having a more diverse workforce. This involves making sure that recruitment procedures encourage applications from, and the actual hiring of, men and women from a variety of backgrounds. Everyone should have an equal chance to work either on-site or in management roles. People from different backgrounds bring refreshing new ideas about design, problem-solving and other workplace issues.

Team roles

The team members on a typical large-scale construction project and their roles and responsibilities are as follows:

The client

The client is the person or organisation at the top of the pyramid – the ultimate employer. It is their project that everyone else is working on. The client may take an active role or **delegate** all or some of the decisions to a contractor. The client usually selects the team members, by appointment or by **tender**. Team members may be selected on the basis of their:

* reputation, expertise and quality of work

* being local to the area, therefore good for the local economy

* history of working with other members of the team before

* price.

The architect

The architect must first consult very closely with the client about the design of the project, and then make sure that all aspects of the design brief are communicated clearly to all members of the team. For many large projects the architect will create a scale model to show what the finished structure will look like. Throughout the construction phase, the architect will:

* consult with all team members on the fine detail of the design and specifications

* assist with problem-solving as issues arise.

ASK Interview three different construction project team members, and for each person find out:

* the main aspects of their role and responsibilities

* the people they interact with on the project (note down the reasons and how often)

* what makes them feel valued and how they value others.

Consulting engineers

The consulting engineer will clarify specifications with the client, architect, surveyor and building contractor to make sure that all the details are correct, suitable and understood by the team members. The specifications may be changed as the consulting engineer collects information on-site. The consulting engineer will oversee the work on-site, dealing directly with the project or site manager and all members of the construction team.

Architectural technologists

Architectural technologists work on-site more than the architect, dealing with team members on the development of the project. The technologist will gather information about particular design issues and pass this on to the team, assisting with problem-solving as necessary.

The information gathering could include:

* clarifying the client's preferences on details of the design

* researching:
  – legal requirements relating to the design
  – selection of materials

* feedback from the client and stakeholders on:
  – the progress of the project
  – the quality and suitability of the final structure.

Project manager

Also called the site manager, the project manager is responsible for making sure that the construction team 'pulls together' and the project is completed safely within the planned timeframe and budget. The responsibilities include:

* liaising with all team members

* planning work schedules

* supervising site preparation

* briefing the workforce

* monitoring progress

* reporting to the client.

Estimator

The estimator works out the cost of the project, including materials, equipment, transport and labour. The project may be divided into components, which can be tendered out. For example the client may invite tenders for the plumbing component or the electrical component of the project. The estimator works out the specifications for the tender, consulting with other team members such as the buyer, surveyor and project manager. When the tenders come in they are compared to the estimator's costing.

Safety officer

The site safety officer is responsible for making sure that all site health and safety requirements are met. The safety officer communicates with all site personnel about specific precautions and safety rules, and checks that these are followed. If an accident occurs, the safety officer conducts an investigation to find the cause, and will make changes to safety procedures to ensure it doesn't happen again.

Surveyor

The surveyor's role is similar to a quality checker. The surveyor makes detailed inspections of the project at various stages and reports the findings to the project team. The aim of the inspections is to:

* check that the construction is structurally sound

* identify defects, faults and quality issues

* ensure compliance with: building regulations, fire-safety requirements, accessibility specifications, sustainability standards and other quality standards specified by the client.

Where problems are identified, the surveyor will work through the issues with the relevant team members to ensure a suitable solution.

Clerk of works

Also known as a site inspector, the clerk of works makes sure that the work on-site is completed to specifications. The site inspector will also keep records of on-site activities, people and events including:

* regular site personnel – numbers and job titles

* site visitors

* deliveries

* documentation received, such as: drawings, specifications and instructions, especially where changes have been made to the original plan

* weather and any other relevant events.

Craftspeople

Craftspeople usually work within their own team on large projects. The team leader will inform the main construction team about design, planning, materials and other information. Each member of the craft team is responsible for:

* the work being completed to the required standards, including safety and quality

* reporting any issues, problems or faults – this is very important.

Subcontractor

The subcontractors are brought in to do particular jobs, such as the plumbing or electrics. Contractors should be included in team notices, briefings and other communications. They should also be committed to the goals of the project and to open communication regarding mistakes, faults, breakdowns and other issues.

THINK?

Imagine you owned a piece of land just outside a small town and decided to build six three-bedroomed houses on the site. Think about the people you would recruit for the project team. List the responsibilities of each member.

Remember: the solutions to any problems that arise must be suitable for the whole project, not just one section.

BRICK TALK

27

ON FILE

TWO DIFFERENT WAYS TO ORGANISE A TEAM

Each project has an organisational structure or framework that shows who manages what, and the lines of communication between team members. The structure ensures that communications are managed efficiently. If everyone on-site called the client or project manager every time there was a problem, nothing would ever get done!

FIG 1.2: Team structures

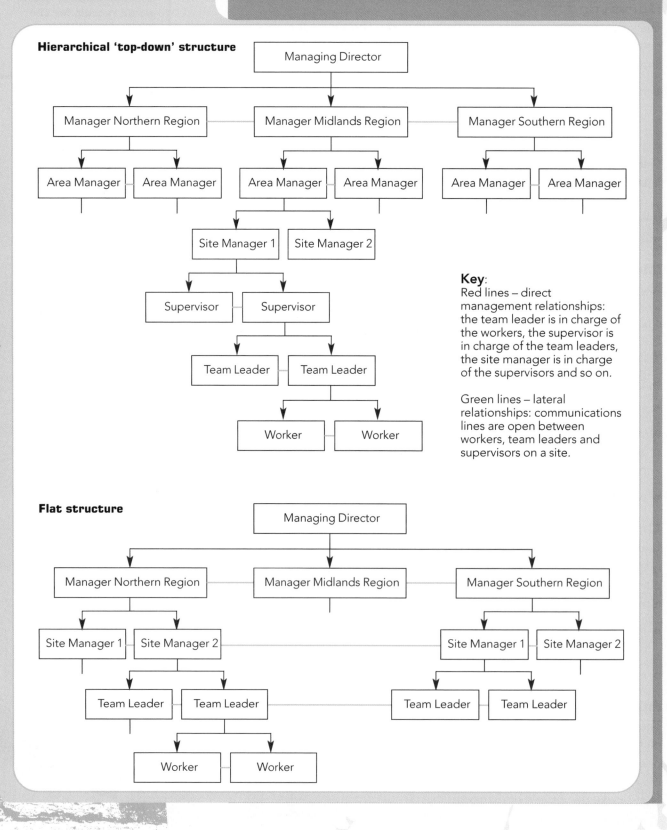

Hierarchical 'top-down' structure

Key:
Red lines – direct management relationships: the team leader is in charge of the workers, the supervisor is in charge of the team leaders, the site manager is in charge of the supervisors and so on.

Green lines – lateral relationships: communications lines are open between workers, team leaders and supervisors on a site.

Flat structure

High-rise apartments: construction project

Boss Buildings is a large company that won the tender for this project. The company employs regional and area managers, site managers, engineers, architects, surveyors, and a team of skilled craftspeople and technicians. The construction team working on this project was managed using a 'top-down' structure. In this structure, the control is held by the people at the head of the organisation. There is some delegation, but most plans, changes and solutions to problems generally come 'from the top', that is, from the bosses. There may be several layers of management in between the boss and the workers. Halfway through the project, Jackie, one of the site managers, returned from a management course and proposed that the team structure be changed to a 'flat' structure. In this structure, there are fewer layers of management and more delegation of responsibilities. Jackie said: 'Communication is faster and more efficient in this type of structure, and teams are in closer contact with each other and may share the load of the work. Also, you can build a positive workplace culture faster, there is more trust'.

TASK

1. The 'top-down' structure is less popular these days: studies have found that employees prefer to be involved in decision-making. Do you agree? What effect does it have on employees if they are involved in making decisions about their work?

2. Mo, a painter, turns up for work one day to find his colleague is off sick. He has a friend, also a painter, who is available for work at the moment. The site manager has to approve replacement staff arrangements. Look at the organisational chart for the two different structures and work out how Mo could get permission to get his friend over for the day to help out. Which structure enables the fastest communication?

3. Can you think of any drawbacks of having a flatter structure, with more responsibility being taken by everyone?

NEED TO KNOW...

GO TO...

For information on the web about **diversity in the construction industry** and **organisational structures** see the resources on the CD accompanying this book.

HUH?

Brownfield site – an urban site that has been built on before

Tender – offer a job for people/companies to bid for

Delegate – pass the responsibility for a task or job to another person or organisation

TOP TIPS

✱ If your organisation includes workers in decision-making – get involved

✱ Ask your colleagues for help and advice

CHECKED

CHAPTER ONE	An introduction to the construction industry	BTEC FIRST
SUBJECT	Construction project teams and teamwork	

➔ Construction team members must: understand the goals and timelines of the project, work efficiently and to a high standard, communicate honestly and openly, especially about errors and problems, and trust and respect the client and other team members

➔ Communication about project progress should be regular, frequent, clear and simple

DIFFICULTY RATING	DATE

Career paths, training and education

The construction industry offers a wide variety of training and education leading to a range of careers. Almost every professional you talk to in the industry will have a completely different story about how they reached their current position. Some people learn mainly on the job, with part-time study, starting as a general operative and progressing up through the qualification levels. Others opt to study full-time and go for a higher qualification straight off. There are many other options in between.

Keep on learning

Successful people in the construction industry build their careers by continuing their **professional development**. Joining an industry group, reading industry magazines and attending short courses, workshops and seminars keeps you up to date with new developments, new technologies and new strategies.

Whichever starting point you choose for your career, your future opportunities are almost limitless. You may be a general operative today and a consulting structural engineer or architect in the future. You may start as a bricklayer, and end up being project manager of a multi-million-pound construction project overseas. You could start as a trainee quantity surveyor and end up running your own successful national business. But you don't need to move from job type to job type if you don't want to. You can stick with your chosen craft or service and become a master craftsman or a specialist in your particular field. Skilled and reliable workers are highly valued across the industry. The point is, the choice is yours, and the scope in the construction industry is almost limitless.

The training for the different levels of occupation is outlined below.

General operative

Some people choose to start full-time paid work as early as possible, and in the construction industry this would mean starting as an entry-level general operative. The training requirements are NVQs/SVQs in Construction and Civil Engineering Services (Construction Operations) Levels 1 and 2. A general operative may decide to apply for an **apprenticeship**, or may study part-time to gain higher qualifications and progress through the company.

Craft occupations

For craft and similar-level occupations, most people start with an apprenticeship. These are available in construction (craft), electrical and electronic servicing, mechanical engineering services and plumbing. An apprenticeship is half training course, half work. You apply for an apprenticeship just like any other job. You can start any time from your 16th birthday up to your 25th birthday. Depending on the workplace, the training required, the training providers available and the type of apprenticeship, it could take between one and five years to complete. A fully qualified craftsperson may work alone, or supervise one or more general operatives, and on large projects may work under the direction of a technician or professional.

There are many craft occupations to choose from

Technical occupations

Some people in technical occupations start by completing an advanced or modern apprenticeship, such as the National Apprenticeship Scheme for Engineering Construction (NASEC). Other people choose to work part-time and attend college to complete qualifications such as a foundation degree; BTEC Certificate, Diploma, Higher National Certificate or Higher National Diploma; or a City & Guilds Certificate. Technicians may supervise craftspeople and work under the direction of a fully qualified professional, such as an engineer or architect.

Professional occupations

For professional roles in the industry, such as an architect or consultant engineer, you need a degree as the first step. After your degree, you may need to do one or two years of supervised work experience, a further degree course, more work experience, and then take a professional exam. It takes at least seven years to qualify as an architect and become a member of the Royal Institute of British Architects (RIBA). An engineer could start with a first degree, foundation degree or a BTEC HNC/HND. Graduate apprenticeship schemes are also available. Professional status as an incorporated engineer requires work-based study, professional development and a professional exam. To be a chartered engineer you need to complete a **masters degree**, then gain work experience and pass a professional exam.

Learn something new every day and keep ahead!

BRICK TALK

CHECK IT OUT

Visit careers offices or websites to gather information and then list as many types of career as possible within the construction industry under the headings: professional, technical, craft, operative.

Choose one occupation from each group and find out what the training options are.

THINK?

Which do you think is best for you?

1. To work your way up from operative to craft career, then technical then professional?

2. To go directly to professional training?

What are the pros and cons of each option? What are the reasons for your choice?

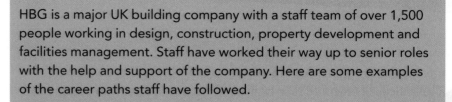
ON FILE

CAREER PATHS AT HBG

HBG is a major UK building company with a staff team of over 1,500 people working in design, construction, property development and facilities management. Staff have worked their way up to senior roles with the help and support of the company. Here are some examples of the career paths staff have followed.

Site Manager: I joined HBG at 16 as a General Building Operative on the Youth Training Scheme and worked my way up to become Site Manager. The company trained me and kept helping me to move up the career ladder. Now I am a member of the Chartered Institute of Building.

Assistant Site Manager: I always wanted to be a Site Manager, and hope to become a Project Manager. I did a year-long placement with HBG as part of my degree course in Construction Management. HBG sponsored me through my final year and I've joined HBG as a graduate Assistant Site Manager.

ASK
Why is professional development important?

Survey at least three experienced construction workers and ask them about the importance of professional development.

Find out what professional development they participate in themselves.

What professional development could they recommend for you?

Project Surveyor: As part of my full-time degree in Commercial Management and Quantity Surveying, I did a one-year placement with HBG. HBG sponsored me through my final year of studies and I returned to work for them when I finished. Almost straight away, I had the chance to run my own project – a great experience.

Senior Project Surveyor: I left school at 16 and worked as a construction trainee four days a week while doing day-release NC and HNC qualifications in Building Studies. With HBG, I applied for and completed a degree in Quantity Surveying – on day release. By the age of 23 I had a degree, no debt and seven years' valuable construction experience.

TASK

1. Compare the career paths of the Assistant Site Manager and the Site Manager. Which career path would you prefer to take? Explain the reasons for your choice.

2. What are the benefits and drawbacks of studying while working part-time?

3. In what way can companies help to finance the further education of employees?

4. What are the benefits and drawbacks of studying full-time, and then applying for a job?

NEED TO KNOW...

GO TO...

For information on the Web about **careers** and **career progression in the construction industry** see the resources on the CD accompanying this book.

HUH?

Professional development – progressing your career by increasing your knowledge through courses, workshops etc.

Apprenticeship – a scheme to learn a trade or skill by a combination of paid work and part-time study

Masters degree – a high-level degree that follows after a first or bachelors degree

TOP TIPS

✳ Keep up with new technology, updated legislation and other industry advances

✳ Put time aside for professional development e.g. join an industry group, read industry magazines, go on courses

CHECKED

CHAPTER ONE	An introduction to the construction industry	BTEC FIRST
SUBJECT	Career paths, training and education	

↪ You can study and progress through a construction career in a variety of ways

↪ The minimum qualification to enter the construction industry is NVQ Level 2

↪ Craft and technical jobs require an apprenticeship, HNC or HND

↪ Professional jobs usually require a university degree

DIFFICULTY RATING	DATE

Qualifications and learning

There are many different courses, qualifications and arrangements for study to choose from when selecting career options and professional development in the construction industry.

Accredited courses

Make sure that any course that you sign up for results in an **accredited qualification**. Some organisations offer training related to the construction industry, but the qualification may not be recognised by other key organisations. For example, the course may be equivalent to an accredited qualification, but if the organisation has not been approved to run accredited courses, the certificate itself may prove worthless. You can check that an organisation is approved to run courses and provides accredited qualifications with CITB Construction Skills, the national training organisation for construction in the UK.

Different ways to learn

As well as selecting the right accredited course for your career development, you can choose how you will learn. Here are some of the options.

* **On-the-job**: all apprenticeships, foundation degrees and many other qualifications can be delivered partly in the workplace, where the skills you learn and the tasks you do at work are planned by your supervisor as part of your training. These courses also have an off-the-job component where more formal training is delivered at a college or university, by **day release** or **block release** from the workplace.

* **At college**: you can choose to attend college full-time to get all your formal basic training completed in one go.

* **Open learning** or **distance learning**: this is a flexible approach to learning, where you complete any practical activities and written assignments at your own pace. Students get online, email or phone support from tutors.

* **online learning**: this refers to courses run on the internet. You can register for the course, view and download materials and complete assignments online. This often has the advantage of letting you complete courses at your own speed. You can also often start a course at any time – you don't need to wait for the beginning of the school or college year in September.

FACT IS...

Here is a list of the technical certificates available and their equivalent NVQ level:

BTEC Introductory Diploma – NVQ Level 1

BTEC First Diploma – NVQ Level 2

BTEC National Award – NVQ Level 3 (1 'A' level)

Ordinary National Certificate (ONC) – NVQ Level 3 (2 'A' levels)

Ordinary National Diploma (OND) – NVQ Level 3 (3 'A' levels)

Higher National Certificate (HNC) – NVQ Level 4

Higher National Diploma (HND) – NVQ Level 5

[LOW DOWN]

Construction Skills Certification Scheme (CSCS) The CSCS card lists the relevant training a person has completed and certifies that they are competent in their job and are health-and-safety conscious. The card also acts as an identity card. The scheme started ten years ago and there are now over 800,000 card-holders in the industry.

CORGI registration CORGI is a register of over 50,000 gas installation businesses with qualified tradespeople, including gas installers and plumbers. The tradespeople must have recognised qualifications and are also assessed for competence before being registered.

You can study at home with distance or open learning courses

CHECK IT OUT

Which sector of the construction industry appeals to you the most?

Find out how many different types of course there are which relate to your chosen area. How many of these include on-the-job training?

Bridging courses

Entry into higher education courses usually requires 'A' levels or equivalent qualifications. However, if you don't have these, there are many bridging courses available to people with apprenticeships or extensive work experience that provide a better route into higher education courses. For example, entry into HNC Engineering/ Construction/Civil Engineering usually requires a National Certificate or Diploma, but there is also an HNC bridging course available.

A person with a Higher National Diploma is already part-way to completing a degree. For example, design students with a good HND foundation degree, or equivalent, in design can complete a bridging course in the summer vacation and start Level 3 of a degree course, graduating after about 11 months of study.

ASK What licences are needed in your chosen career area?

What is the best way for a person in your chosen career area to show that they have the proper training and are and completely compete reliable?

35

Qualifications for craft and technical occupations

National Vocational Qualifications (NVQs) (SVQs in Scotland)

These qualifications are based on the National Occupational Standards. The trainee must be assessed as competent before the qualification is awarded and the trainee moves to the next level. The NVQ levels apply to workplace levels of responsibility as follows:

* Level 1 – performs a range of routine and predictable work activities.

* Level 2 – performs a wide range of work activities; some tasks will be complex or non-routine and/or unsupervised; the person often works as part of a team.

* Level 3 – performs a broad range of complex and non-routine work activities, taking responsibility for the work and often controlling or guiding others.

* Level 4 – performs a broad range of complex, technical or professional work activities, taking responsibility for a part of a project.

* Level 5 – takes responsibility for projects, including analysis, diagnosis, design, planning, execution and evaluation.

At each level, a specified number of units of training must be completed.

Apprenticeships

Most apprenticeship are made up of: key skills training, providing foundation employability skills; relevant NVQs/**SVQ**s; and in some cases, a technical certificate. The NVQ component of apprenticeship training is as follows:

NVQ Level 1   – vocational access training (Northern Ireland)

NVQ/SVQ Level 2 – traineeship (NI)
                               – foundation apprenticeship
                               – specialist/civil apprenticeship (Scotland)

NVQ/SVQ Level 3 – advanced apprenticeship
                               – modern apprenticeship.

Technical certificates

Technical certificates also form part of some apprenticeships, and the higher-level certificates can form part of a degree qualification.
See *Fact is* on page 34.

Professional qualifications

Professional qualifications include:

* Foundation degrees: Study while you work! Foundation degrees in construction, such as Sustainable Construction, Quantity Surveying, Interior Architecture, and Built Environment (Construction) are delivered in partnership with employers. You need an NVQ Level 3 to start a foundation degree, or you can apply with two years' relevant work experience, then complete an orientation course (similar to a bridging course).

* First or undergraduate degrees: Three-year full-time courses, longer if work placements are included, such as BEng (Bachelor of Engineering).

* Masters degrees: An extra year of advanced study, either by tuition followed by project work, or by approved research.

You can take a number of different pathways to achieve a degree qualification

NEED TO KNOW...

GO TO...

For information on the web about **CORGI**, **CSCS**, **apprenticeships**, **accredited courses** and **foundation degrees** see the resources on the CD accompanying this book.

HUH?

Accredited qualification – a qualification issued by an approved organisation is recognised by the industry

Day release – the release of students from work for a day at a time to attend college

Block release – the release of students from work for blocks of time, such as a week or a term, to attend college

SVQ – Scottish Vocational Qualification. The equivalent to an NVQ in Scotland. It is a very similar qualification to an NVQ

CHECKED

CHAPTER	An introduction to the	BTEC
ONE	construction industry	FIRST
SUBJECT	Qualifications and learning	

➲ **Apprenticeships include: key skills, NVQs or SVQs, and sometimes a technical certificate**

➲ **Technical certificates include: BTEC Introductory Diploma, BTEC First Diploma, BTEC National Award, ONC, OND, HNC and HND**

➲ **Professional qualifications include: foundation degrees, first (undergraduate) degrees, masters degrees**

➲ **Studying methods include: learning on the job, attendance at college, open or distance learning**

DIFFICULTY RATING	DATE

Construction projects: stages of development and influencing factors

We have looked at the wide variety of construction projects and the people who work on construction teams – now we are going to explore the main stages of construction projects: project planning, design, construction planning and actual construction. At each stage there are several significant factors to consider, such as physical, technical, financial, legal, environmental and **aesthetic** issues.

Project planning

The first step in planning a construction project is to define the scope – that is, exactly what the project will involve and what the deliverable products are. This sets the limits of the project, so that everyone is clear about what the client expects and the timeframe for completion. The legal contract will be based on this.

The details of the client's requirements must be defined and any legislation, regulations or standards must be considered. For example, are there restrictions on planning permission, such as land use, or height of buildings; are there any **heritage factors** to be considered; is an **environmental impact assessment** required? The local community, users of the finished project and any other interested parties should also be consulted, to ensure that the project is in harmony with the local community and environment.

> In construction, problem-solving is all part of the job!

BRICK TALK

A public meeting may be held to discuss the proposed plans and designs

[LOW DOWN]

→ **Aesthetic influences** Does the structure add to the landscape and improve the view? Is the design attractive to the users? Does the design block the view of any other buildings or cast a shadow?

→ **Environmental impact assessment (EIA)** A full analysis of the potential environmental effects of a construction project locally and on a wider scale, depending on the project. The people conducting the EIA look at the possible effects on water, air, noise, traffic and transport, landscape, urban design and wildlife.

→ **Heritage factors** These are concerns about the history of the area or site. Construction projects need to be sensitive to historical remains so that features of historical or archaeological value are not destroyed by new projects. A new construction may need to be in keeping architecturally with its surroundings.

Design

The design stage involves extensive negotiations with the client, especially where legislation or other matters conflict with the client's requirements. There may also be open consultation with the local community for the design to be discussed, issues identified and solutions negotiated. The design phase must take into account the aesthetic appeal of the structure, both inside and out, and the effect of the structure on the environment. The design should also be within the **budget** set by the client.

Construction planning

When the design is finalised, and specifications prepared, detailed planning of the construction phase can begin. This involves breaking down each section of the work, and defining team assignments and timelines. Material requirements must also be planned, and estimates and costings prepared. Risk assessments and risk management strategies should also be planned.

Before construction can begin, the correct forms of authority to build must be obtained from the local authority and any other legal permissions obtained, for example concerning traffic management to and from the site.

CHECK IT OUT

What are the important factors to consider when planning work in your chosen area of the construction industry? Consider the factors we have discussed in this topic: aesthetic, environmental, financial, legal, physical, technical.

ASK Find out what experienced construction workers think of the lean construction methods. Talk to at least three experienced construction workers and ask how they would improve work planning to be more efficient.

Actual construction

While construction is actually going on, the site will be extremely busy with the movement of personnel and materials; this is the stage when the design becomes reality, and there are usually 'glitches' – no construction project ever runs completely smoothly and exactly according to plan.

Some of the main factors that arise during the construction phase are as follows:

* **Physical** – such as maintaining health, safety and welfare systems on-site and at the site boundaries; ensuring timely ordering and supply of human and material resources.

* **Technical** – such as ensuring that the design and specifications are interpreted correctly and followed; ironing out problems that occur due to unforeseen circumstances.

* **Financial** – such as keeping to estimated budgets; maximising the efficient use of time and materials; controlling waste and energy use.

* **Legal** – such as making sure that solutions to problems comply with legal requirements; ensuring that the correct safety inspections, **health surveillance** and building inspections are conducted to schedule.

* **Aesthetic** – such as keeping the site surround fencing in good condition and cleaning up waste and debris on access routes.

* **Impacts on the natural environment** – such as clearing away waste on-site promptly; never burning waste on-site.

Keep access around construction sites clean, tidy and safe for pedestrians

ON FILE

LEAN CONSTRUCTION

Lean construction is a project planning method designed to reduce waste and inefficiency of all kinds. One organisation used the theory to improve the efficiency of bricklaying processes. The bricklaying team analysed all the steps involved in the laying brickwork, especially re-laying reclaimed facing bricks. The team also investigated access to the work area, work methods, tools and work patterns, and the handling and movement of materials on-site. Analysis of the data collected showed that 45 per cent of the bricklayers' working time was being wasted. A large proportion of this wasted time was caused by waiting for scaffolding installation or excavation operations to be completed. The team leader said: 'Improved interaction and communication with other trades, and detailed planning of the order and timing of the work, should reduce wastage in this area'.

The skill and efficiency of the materials handler also affects the productivity of the bricklayer. Having the materials located close to the work area and **replenished** in good time reduces waiting time for the bricklayer. The team worked out the most efficient method for setting up, operating and controlling brick and block work, and identified the best way of getting the best mortar supply possible: 'We have challenged our habits and really looked at how we do things, and as a result we have improved the productivity of the bricklaying process by seven per cent so far'.

 TASK

1. Consider the main influencing factors on construction projects (physical, technical, financial, legal, environmental, aesthetic). Which of these were involved in the bricklaying team's project to improve planning, and how?

2. Why might the scaffolders or excavators be delayed in completing their work? What should these workers do if they are delayed?

3. Do you think it is a good idea to try to improve productivity to the point where bricklayers can be actually laying bricks 100 per cent of the time during their working hours? Explain your answer.

NEED TO KNOW...

GO TO...

For information on the web about **lean construction and project planning** see the resources on the CD accompanying this book.

HUH?

- **aesthetic** – pleasing to the eye, artistic
- **budget** – the amount of money allocated to pay for a particular job
- **heritage** – of historic value, to be preserved
- **health surveillance** – programme of regular health checks for people working with hazards
- **replenished** – re-stocked

TOP TIPS

✱ Apply lean construction methods to your own work. If you can find ways to be more efficient with your time or materials, you could reduce costs for your employer, and there may even be a bonus in it for you!

CHECKED

CHAPTER ONE	An introduction to the construction industry	BTEC FIRST
SUBJECT	Construction projects: stages of development and influencing factors	

↪ Most construction projects involve the following stages:
project planning
design
construction planning
actual construction

DIFFICULTY RATING	DATE

Sustainable construction techniques

Human beings are damaging the natural environment. Our energy consumption in particular is contributing to global warming. Without a healthy natural environment, there would be no living things on the planet, including humans! It is therefore vitally important that we take steps to protect and respect the natural environment, especially in the construction industry. This is why so many innovative sustainable construction techniques are being developed, such as using alternative energy sources and building 'living roofs'.

The natural environment

You can protect the natural environment at work by respecting your worksite and surroundings, including the air and water nearby. Respecting the natural environment involves all of the following.

* **Reducing waste**: the construction industry produces 90 million tonnes of construction and demolition waste every year – landfill sites are almost full, so where will it all go?

* **Reducing pollution**: the construction industry is responsible for a third of all industry-related pollution incidents.

* **Controlling consumption**: the construction industry accounts for 10 per cent of the national energy consumption – the production of energy produces greenhouse gases, so the less energy we use, the less we need to produce – reducing greenhouse gases.

* **Conserving natural assets**: the construction industry builds on 6,500 hectares of rural land each year – if this carries on there will be no countryside left!

* **Preserving wildlife**: ecosystems are very finely balanced systems of local plants and animals – everything works together well, because the system has developed over hundreds of thousands of years. An imbalance in the system, such as loss of habitat for animals, can have devastating effects.

FACT IS...

Respecting the natural environment involves:

* sorting and recycling waste materials
* minimising pollution
* reducing energy consumption
* conserving natural assets (rural sites)
* preserving wildlife
* protecting biodiversity
* using alternative energy sources.

[LOW DOWN]

→ **Brownfield site** Any land (with or without buildings) which was previously used, but is not used now – the land may be vacant, derelict or contaminated – quite often the land must be decontaminated and restored before being re-used. The Millennium Dome was built on a contaminated brownfield site.

→ **Biodiversity (biological diversity)** The range of different types of life form, plant and animal, which co-exist and interact in an area.

A house with a living roof

CHECK IT OUT

What are the environmental problems in your local area?

Are the waste tips full?

What are the pollution levels – on the land, in the water, in the air?

Are any local ecosystems under threat?

Is any local wildlife in danger of losing its habitat?

Sustainable construction techniques

There are several innovative techniques that the construction industry can use to reduce the impact of construction processes and products on the environment.

* **Environmentally friendly design** – such as:
  - passive heating – the location and alignment of a building, placement of windows and careful selection of materials can make the best use of 'passive heat', that is, the warming effect of the sun; this will reduce energy consumption
  - **living roofs** – which have soil and plants on them improve air quality, reduce stormwater run-off into rivers, and add to the insulation of the building, reducing energy consumption. They can also provide habitats for birds and animals.

* **Specification of locally sourced and sustainable materials** – this reduces transport costs and pollution and supports the local economy; sustainable materials are those made from renewable resources, using alternative energy sources – for example timber from a sustainable forestry plantation milled using wind power.

* **Improved site and resource management** – there is a move towards lean construction, which involves working out ways to reduce waste of all types during a construction project, including wasted time, energy, resources and materials.

ASK What construction materials or products are processed or manufactured in your local area?

Are the materials produced using sustainability principles?

How much of the materials is used locally, reducing transport costs, fuel consumption and pollution – and how much is delivered to distant locations?

Use separate skips for disposal and recycling of different types of waste on-site

* **Waste management, reclamation and recycling** – involves sorting waste on-site into separate skips, for example paper, timber and brick; and re-using materials wherever possible, or using products made from recycled materials.

* **Alternative energy technology** – designing buildings with solar or **biomass** energy, and heat pumps to reduce energy taken off the grid – this will gradually reduce the requirement for national energy production in power plants and therefore reduce greenhouse gases.

* **Use of brownfield sites** – reducing the amount of building on rural sites and increasing construction on brownfield sites (pre-used sites that may have been contaminated by the previous building and activities). The government intends to ensure that 60 per cent of new development occurs on brownfield sites.

Modern homes are being designed more and more with the environment and energy efficiency in mind, like these solar panels

ON FILE

THE GENESIS CENTRE

The Genesis Centre is a demonstration building forming part of the Genesis project at Somerset College of Arts and Technology. The Centre is a showcase for sustainable construction and has a number of 'pavilions' built using different techniques – earth, straw, clay and timber. Each of the pavilions has a living roof. The water pavilion (toilet) demonstrates the latest water-saving technologies. The internal sections of the pavilion are exposed to show the construction technique, and each one is monitored for performance against industry standards. The drainage system uses landscaping to slow down the water run-off and filter it naturally, creating wetlands and ponds. The Centre uses solar and **biomass** energy – the fuel for the boiler is compacted wood dust from the carpentry workshops.

The Genesis Centre is dedicated to sustainable construction techniques

TASK

Explore the Genesis Centre website, and find out all the detailed information about the 'pavilions' http://www.genesisproject.com/genesis05/ then answer the following questions:

1. Choose two types of structure used for the pavilions and describe the techniques used.

2. Which pavilion do you think will be the most durable – able to withstand weather and wear and tear? Note down the reasons for your selection and the reasons why you think the others are less durable.

3. What is biomass energy? What other types of fuel could be used?

4. What are the benefits of a living roof?

NEED TO KNOW...

GO TO...

For information on the web about **the impact of the construction industry on the environment**, **sustainable construction** and **lean construction** see the resources on the CD accompanying this book.

HUH?

Biomass – the total amount of living material (plants and animals) in an area. Biomass energy uses organic fuel, such as wood

TOP TIPS

* Think twice before you put anything in the general waste bin or skip to be taken to landfill. Can the item be re-used or recycled?

* Some used products and materials are worth money and can be sold

CHECKED

CHAPTER ONE	An introduction to the construction industry	BTEC FIRST
SUBJECT	Sustainable construction techniques	

- An eco-system can take a million years to develop and only one day to destroy

- Construction companies should use environmentally-aware procedures and products wherever possible

- Environmentally friendly design includes: passive or solar heating, living roofs, grey-water systems, recycled or sustainable materials

DIFFICULTY RATING	DATE

Unit summary

✔ Creating a sustainable built environment

Significant features of the built environment include: location, height and shape, access, open space, energy requirements, suitability for purpose. Creating a sustainable built environment includes:

* making sure air and water are kept clean
* leaving plenty of natural open spaces
* not over-using limited resources
* not damaging the planet beyond repair
* careful planning, design and construction
* environmental impact surveys
* consulting all stakeholders
* waste minimisation.

✔ The construction industry is important – socially and economically

The construction industry supplies communities with houses, workplaces, schools, hospitals, cultural and leisure centres and shopping centres, and the transport, communications, water, waste and power infrastructures that service these buildings. Economically, the construction industry provides jobs for about three million people, some working in the industry and some extracting materials, manufacturing supplies or providing services used by the industry.

✔ What's going on in the construction industry?

There are many different job activities in the construction industry, including:

* planning – what goes where?
* architecture – what will it look like?
* surveying – does it meet legal requirements?
* civil and structural engineering – will it stay up?
* building crafts – making the design real
* building services engineering – does it have lighting, heating and cooling?
* estate management – what needs to be done?
* facilities management – is it running smoothly?

✔ The wide range of construction projects

The range of construction sectors includes: residential, commercial, retail, cultural, leisure, industrial, health, educational, agricultural, utilities and services, public buildings and transport infrastructure. Within each of these sectors, there is a range of activities, such as planning and design, new build, restoration and renovation, and maintenance and repair. Clients include: private individuals, sole traders, private companies, public limited companies and governments.

Construction project teams and teamwork

The project team may include: the client, architects, architectural technologists, surveyors, clerks of works, project managers, skill team managers or leaders, safety officers, craftspeople, general operatives, estimators, buyers, consulting engineers and subcontractors. As well as the relevant skills, team members must have an understanding of goals and timelines, a commitment to quality standards, a commitment to open communication, and trust in and respect for the client and other members of the team.

Career paths, training and education

There is a wide variety of careers and training and education options. Professionals in the industry reach their positions by a range of routes – some learn on the job with part-time study, others opt to study full-time and go for a higher qualification straight off. There are many other options in between.

Qualifications and learning

There are many different courses, qualifications and arrangements for study to choose from when selecting career options and professional development in the construction industry. Make sure that courses are accredited. Qualifications include National Vocational Qualifications (NVQs – SVQs in Scotland); apprenticeships; technical certificates, such as Higher National Certificate (HNC); Higher National Diploma (HND) and professional qualifications such as undergraduate degrees and masters degrees.

Construction projects: stages of development and influencing factors

The three main stages of construction projects are: planning (project planning and construction planning), design and actual construction. At each stage there are several significant factors to consider, such as physical, technical, financial, legal, environmental and aesthetic.

Sustainable construction techniques

Respecting the natural environment involves sorting and recycling waste materials, minimising pollution, reducing energy consumption, conserving rural sites, preserving wildlife, protecting biodiversity and using alternative energy sources.

When you have completed this unit, you should be able to:

✓ explain the importance of health, safety and welfare in the construction industry, including:
- the elements and main causes of accidents
- the human and financial costs of accidents

✓ describe your personal responsibility for health, safety and welfare in the construction and built environment sector, including:
- responsibilities under the Health and Safety at Work Act
- risk assessments; safety policies and safe systems

✓ identify and describe the hazards and risks you would expect on construction sites, including:
- identification of hazards and risks that could arise when using plant, equipment, machinery and materials
- factors that affect the level of risk, such as: workplace factors, human factors

✓ describe good methods for ensuring the health, safety and welfare of workers including:
- regular training, good workplace procedures, provision of appropriate personal protective equipment (PPE)
- appropriate fire precautions and training, workplace signage and good housekeeping
- awareness of safe work methods for working at heights, working below ground and working in confined spaces
- compliance with the Provision and Use of Work Equipment Regulations (PUWER)
- training and appropriate workplace procedures for using electricity on site and being aware of buried and overhead services.

Exploring health, safety and welfare in the construction industry

Scale: 1:1

Coming up in this unit...

Hazards are present in almost all workplaces but, because of the type of work, construction sites are much more dangerous than most other work locations. Accidents can cause damage to property and equipment and, more importantly, injury to workers, site visitors and the general public. Both employers and employees can be fined if it is proved that their action – or inaction – contributed to an accident. To reduce the likelihood of accidents, day-to-day workplace health, safety and welfare procedures and best-practice control methods must be followed at all times. Risk assessments should be carried out both for the workplace and for specific tasks, and their recommendations carried out.

Some accidents are fatal; in these cases many people are affected.

It is your responsibility to keep yourself, your colleagues and members of the public safe and well through your actions at work. To do this, you need to develop a good foundation of knowledge and skills about health, safety and welfare. Remember – you will never stop learning and updating your knowledge as you progress through your career. New circumstances and technologies will call for new work methods and techniques, and these will require new strategies for risk control. You need to pay at least as much attention to safe methods of doing your work as you do to your specialist work skills.

Keep yourself and others safe and well at work

Avoid accidents!

Accidents will happen and, unfortunately, some of those accidents will be fatal. On average, more than 60 people are killed every year in the UK through accidents on construction sites. Many more people suffer permanent injury, disability or the loss of an arm or leg. But you can do a lot to avoid accidents. You take care not to get killed or seriously injured in other situations – like crossing a busy road – so surely you should take care not to have an accident at work either. Makes sense, doesn't it?

Accidents in the construction industry

An accident could be:

* a fall
* a slip or trip
* injury caused by tools and equipment
* injury caused by materials
* poisoning
* violence inflicted by work colleagues.

Accidents are caused by:

* unsafe conditions – such as:
  – faulty or inadequate equipment
  – the presence of hazardous substances
* unsafe acts – mistakes made by people, such as:
  – failing to wear goggles or a hard hat
  – failing to follow safe work methods.

What is this worker doing wrong? Over half of all fatal accidents in the construction industry are falls from height

The majority of accidents in the construction industry are slips, trips and falls; these cause over 1,000 long-term injuries every year. The most frequent type of fatal or serious accident on construction sites is falling.

FACT IS...

The most common serious accidents are caused by the following:

Falling from a height, such as:

❋ through a fragile roof

❋ through a skylight

❋ from a ladder

❋ from a scaffold.

Being hit with force, such as by:

❋ moving vehicles and plant, e.g. excavators

❋ falling loads, e.g. bricks or girders

❋ falling equipment, e.g. power tools

❋ collapsing structures.

[LOW DOWN]

→ **Accident** An accident is an unplanned event that results in the injury or ill health of people, or damage or loss to property, plant, materials or the environment, or a loss of business opportunity.

→ **Near miss or close call** More formally called a 'dangerous occurrence', a 'near miss', or 'close call' is an unplanned event that does not result in injury, illness or damage – but very easily could have.

→ **Chain of events** The chain of events includes all the things that happened in the lead-up to an accident and which were definitely connected with that accident. The difference between an accident and a near miss is often just one small difference in the chain of events.

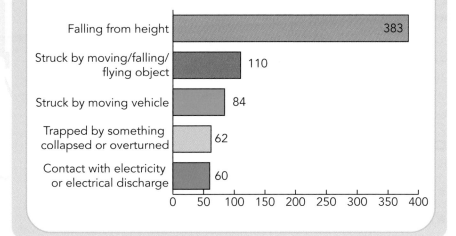

FIG 2.1: Don't become one of the statistics!

Number of fatal injuries to workers in construction 1996–2006

Falling from height	383
Struck by moving/falling/flying object	110
Struck by moving vehicle	84
Trapped by something collapsed or overturned	62
Contact with electricity or electrical discharge	60

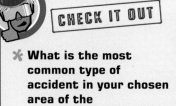

Over the last five years, the rate of fatal injury has fallen by 50 per cent. Keep up the good work!

BRICK TALK

Elements of an accident

The elements of an accident include any factors that contribute to the accident. These may be:

* **organisational**: management decisions, project timelines

* **technical**: suitability of tools, equipment or materials; **environmental** influences such as light and heat

* **human error**: lack of concentration, stress, working above your skill level, not reading instructions carefully, taking short cuts, untidiness.

Accident investigation

Accidents are investigated to discover:

* the root cause

* the contributing elements

* the chain of events.

This information is used to decide what action to take to prevent such an accident happening again. This might include changing or improving procedures and training, or getting better equipment.

A 'near miss' or dangerous occurrence should also be investigated. These are cases where an accident is only just avoided. The chain of events and root cause could have led to an accident, and may do next time, so it is just as important to look into these events. The difference between a near miss and an accident may be just one small factor, such as timing. An example might be a crane-load slipping and falling and narrowly missing a person walking underneath. In this case, at least two elements require urgent investigation:

1. Why did the load slip?

2. Why was the person walking underneath the crane?

CHECK IT OUT

✳ **What is the most common type of accident in your chosen area of the construction industry?**

✳ **What is the most common cause of these types of accident?**

✳ **What steps should be taken to avoid these accidents?**

THINK?

Think about an accident or near miss that has happened to you or someone you know.

✳ **What were the elements of the accident?**

✳ **What was the chain of events leading up to the accident?**

✳ **What was the root cause of the accident?**

The cost of an accident

As well as the enormous human cost of accidents – including injury and even death – accidents can also have a significant financial cost to people, companies and society.

Cost to people

A person injured by an accident can make a claim against the employer for loss of earnings, but this takes time. Meanwhile, sickness benefit paid by the government will usually be significantly less than wages. If the injury prevents the person from returning to work at their original skill level, they may end up on a lower pay rate, or even be unable to return to work at all.

Cost to companies

Financial costs include:

✳ the accident investigation

✳ providing solutions to the accident cause (such as purchase of new equipment)

✳ compensation claims

✳ extended project completion time

✳ recruiting and training of replacement staff.

In addition to financial cost, a company may lose its reputation, which can cause loss of business.

Social cost

This includes the cost of supporting workers who can no longer earn a wage because of illness or disability. There is also the related cost to the whole family of reduced health and well-being and lack of opportunity resulting from having to survive on an extremely low income.

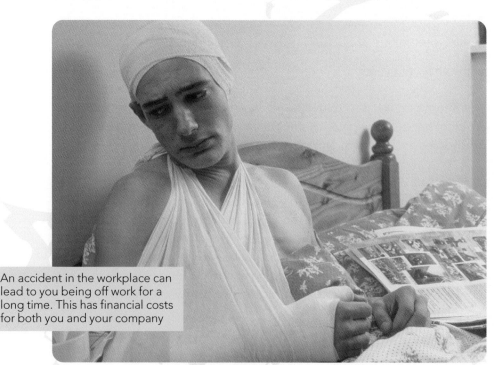

An accident in the workplace can lead to you being off work for a long time. This has financial costs for both you and your company

ON FILE

THE PRICE OF NEGLIGENCE

Jim Baker, a labourer working with a self-employed contractor, had to have his leg amputated below the knee after an on-site accident. Jim was helping the contractor shift materials from the back of a tipper, using a seven-tonne excavator. While Jim was standing on the side of the truck, the contractor manoeuvred the excavator bucket over the load. The bucket suddenly swung, crushing Jim's leg against the side of the truck. The accident investigation found that the contractor had not been trained to use the excavator. He had never attended a Construction Industry Training Board training course, and was not classified as a plant operator. The company employing the contractor was fined £4000 and the self-employed contractor was fined £350.

TASK

1. What were the elements of this accident?
2. What training should a person have before using a seven-tonne excavator?
3. What safety training should site labourers have before helping with jobs such as this?
4. Do you think the fines are fair?
5. What are Jim's prospects for the future?

Take extra care when working in excavations

The law protects you!

The Health and Safety at Work etc. Act 1974 (**HASAWA**) is designed to make workplaces safe and avert all accidents. Under this law both the employer and the employee have responsibilities.

Your responsibilities

It is your responsibility to cooperate with your employer on health and safety issues and to take care of yourself, any equipment you use and any others that may be affected by what you do. This includes:

* following company procedures and work instructions
* using the personal protective clothing and equipment (PPE) required for the job
* working safely, using safe work methods
* using, storing and maintaining tools and equipment correctly
* being aware of hazards and risks and taking recommended measures to avoid them.

Keep yourself alive and well – use the correct PPE and follow instructions

Your employer's responsibilities

The Health and Safety at Work Act states that an employer must take all steps, as far as reasonably practical, to ensure your health, safety and welfare at work, including:

* conducting **risk assessments** of the workplace and of all work tasks
* providing a safe workplace without health risks
* making sure that all plant, machinery, equipment and tools are safe to use

* providing information and training on safe systems of work and ensuring these are followed

* making sure that there are safe systems and procedures for moving, storing and using equipment and materials, including waste

* providing adequate light, heat, ventilation, sanitation, washing, rest, first aid and welfare facilities

* (if there are more than five employees) drawing up a health and safety policy statement and a set of emergency procedures, and appointing an adequate number of safety representatives

* reporting injuries, diseases and dangerous occurrences.

Safety rules

Workplace policies and procedures are written to ensure that work is completed according to legal requirements. Procedures are improved over time, especially using information from accidents and near misses, and updated to comply with changes in technology or the law. Since safety rules and procedures are designed to keep you alive and injury-free, there is every reason to make sure you know what they are and to follow them exactly.

In general, the safety rules on a construction site will include:

* **electrical safety**: knowing the location of underground and overhead cables; using low voltage for tools and equipment; using protected cables, leads, plugs and sockets; regular safety inspection of power tools and equipment

* **emergency procedures**: evacuation plan training and practice; alarms checked regularly; emergency phone available; escape routes identified; first aid available

* **excavations**: planned and inspected by a trained and competent person; adequate shoring safely installed, or sides sloped to a safe angle; safe ladder access; barriers around excavation area; stop blocks to halt vehicles; no weighty materials near edge

* **fire prevention**: minimum quantities of flammable products kept on-site; smoking banned; all sources of ignition banned near flammable store; gas cylinders stored (valves closed) and maintained correctly; combustible waste removed; correct fire extinguishers provided, inspected and maintained

* **hand-arm vibration**: workers trained in effects of over-use of vibrating tools; job rotation and reduced-vibration tools used; regular inspection and maintenance of vibrating tools; health surveillance arranged as required

---

CHECK IT OUT

Scan through articles in the local, regional and national newspapers.

How many reports of workplace accidents can you find?

How many of these happened in the construction industry?

What injuries were there?

What were the penalties?

ASK Find out the detailed safety rules for three different activities in your chosen area of the construction industry, for example:

* using a drill
* using a ladder
* using cement.

Can keeping to the safety rules prevent *all* accidents?

55

A contractor working on-site must ensure that any person employed by them, or any person who is self-employed but under their control:

✳ is trained and competent to do the job safely

✳ is properly supervised

✳ is provided with clear instructions for the work

✳ has access to washing and toilet facilities

✳ has the correct tools, equipment and plant and that these are in good working order

✳ uses the correct protective clothing and equipment.

THINK?

Imagine that you are a health and safety inspector visiting a work site.

What safety procedures would you check?

How would you expect the workers to react to your inspection?

How would you expect the project manager to react to your inspection?

* **hazardous substances**: identified before work commences; risk assessment completed and control measures applied; workers trained in safe use; health surveillance arranged as necessary

* **ladders**: only used if there is no other possible method; ladders should be in excellent condition; rest against a very solid surface; secure to prevent slipping; never stretch to reach work – move the ladder position instead

* **manual handling**: training in safe lifting techniques; equipment provided for lifting heavy objects and materials; materials ordered in, or transferred into, 25 kg bags

* **noise**: reduced wherever possible; training given to workers on effects of noise; correct ear protection supplied

* **safe access to and from work areas**: lighting supplied where necessary; guard rails around or covers over hazardous areas; no obstacles in walkways

* **scaffolds**: installed; inspected and dismantled by trained and competent people; base plates or timber sole plates on uprights; scaffold secured to the building; fully boarded platforms; edge protection on platforms (such as double guard rails and toe boards)

* **tools and machinery**: suitable for the job; guards on sharp edges; moving parts inspected and checked regularly; regular maintenance schedule for all tools and machinery; training provided as required

* **vehicles**: separate road and walkways on-site; warning signs in large plant operation areas; reversing avoided; brakes, lights and steering checked regularly; drivers trained; loads secured

* **welfare**: clean toilets and washing facilities on-site or nearby; drinking water provided; rest area provided; rest periods approved

* **working on roofs**: edge protection provided; safety nets installed correctly (industrial-sized roofs); barriers around fragile roof areas (cement sheets, roof lights); area cleared beneath roof workers.

Well-constructed scaffolding helps to ensure worker safety

ON FILE

ACCEPTABLE RISK?

The **HSE** prosecuted a contractor for placing employees at risk – no accident had occurred. An inspector noticed workers removing slate tiles from the roof of a building that was going to be demolished. The inspector could not see any roof ladders or scaffolding, and stopped to inspect the site. The inspector issued a **prohibition notice** to stop the work and investigated the matter, finding that none of the workers had received safety training in working from height. The contractor said he could not afford to supply ladders and scaffolding as this would reduce the profit from the sale of the recycled tiles. The contractor pleaded guilty to the charges and was fined £3,000.

TASK

1. Is it a good idea for HSE inspectors to visit sites without an appointment to check safety procedures? State your reasons.
2. Why do you think the workers were doing the job, even though it was unsafe?
3. What equipment would you provide for doing the job safely?
4. Should the penalty for placing people at risk of fatal injury be the same as the penalty when a fatal accident has occurred? Explain your answer.

Always follow instructions for staorage or use of hazardous materials

NEED TO KNOW...

GO TO...

For information on the web about **site safety** and **HSE penalties** see the resources on the CD accompanying this book.

HUH?

HASAWA – Health and Safety at Work etc. Act

HSE – Health and Safety Executive

Prohibition notice – a legal order to stop work

Risk assessment – a survey made of a job before anything is done to identify any undue risks to workers or the general public, plus recommendations to minimise those risks

TOP TIPS

✱ Check that the procedures, work instructions and safe-work methods you are using are up to date

✱ Report hazards as soon as they occur

CHECKED

UNIT	Exploring health, safety and welfare in the construction industry	BTEC
TWO		FIRST
SUBJECT	The Law protects you	

➲ **The Health and Safety at Work etc Act. 1974 (HASAWA) sets out safety rules for both employers and employees**

➲ **It is your employer's responsibility to provide a safe workplace**

➲ **It is your responsibility to work safely**

DIFFICULTY RATING	DATE

Hazards and risks

A **hazard** is anything that can cause harm or danger. You need to be alert to the hazards that arise in different situations in the construction industry. A safe work-method for one environment, such as the workshop, may not apply to another, such as a busy construction site in wet weather.

A **risk** is the chance or possibility of danger.

The risk level associated with a hazard is a combination of:

✳ the likelihood of an accident happening

✳ the degree of injury that could be caused.

Risk assessment and control

The Management of Health and Safety at Work Regulations 1999 require all employers to assess the health and safety risks in their areas of responsibility, including risks to contractors and sub-contractors.

A risk assessment identifies the hazards and the likelihood of an accident happening, and recommends control measures to prevent accidents. The main steps of a risk assessment are:

1. Identify the hazards – look at the use of all tools, machinery, equipment and materials, especially chemicals, and include storage, decanting and dilution.

2. Work out who, how and what? – Who uses the equipment, materials, tools or machinery? How do they use it? What possibilities are there for harm to occur?

3. Weigh up the level of risk.

4. Select risk control measures – use the **hierarchy of controls** report.

5. Put risk control measures in place.

6. Monitor the effect of the risk control measures on the level of accidents and incidents.

7. Review the risk assessment regularly.

A hazard is an accident waiting to happen!

BRICK TALK

[LOW DOWN]

➜ **Hierarchy of controls** A system of control selection that starts with the most effective control and, if this is not possible, moves to the next level of control, and so on. The levels are: 1. eliminate the hazard 2. substitute the hazard 3. use engineering controls 4. use administrative controls 5. use personal protective equipment.

Hazardous substances

The Control of Substances Hazardous to Health Regulations 2002 (COSHH) sets out eight basic measures for controlling and assessing the risks of hazardous substances. There are separate regulations for asbestos and lead.

The eight steps recommended by COSHH are:

1. Assess the risks.

2. Decide what precautions are needed.

3. Prevent or adequately control exposure.

4. Ensure that control measures are used and maintained.

5. Monitor the exposure.

6. Carry out appropriate health surveillance.

7. Prepare plans and procedures to deal with accidents, incidents and emergencies.

8. Ensure employees are properly informed, trained and supervised.

Hazardous substances used or found on construction sites include:

✳ concrete

✳ dust

✳ paints

✳ solvents

✳ **mastics**

✳ wood glue and other adhesives

✳ fungicide

✳ fuel

✳ other chemicals.

What are the main hazards associated with work tasks in your chosen area of the construction industry?

What risk control measures are used to prevent accidents occurring as a result of these hazards?

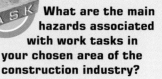
Ventilation and extraction systems must be installed in areas where workers are exposed to hazardous materials

Personal Protective Equipment (PPE)

The risk of a hazard causing an accident can usually be controlled by using the correct safe-work method and wearing the recommended protective clothing and equipment. For example, for most manual tasks there will be a requirement to wear specific types of personal protective equipment, such as goggles or protective gloves.

Hazard category:	Possible accident:
Access	Slips, trips, falls Possibly fatal collision with vehicles
Height	Serious falls Dropping weights on others
Excavation	Being buried under collapsed earth Falling in (people, vehicles or materials)
Manual handling	Serious spine injury Slips and trips
Machinery	Crushed limb (pulled into working part) Serious cuts
Electrical	Electrocution
Fire	Burns Smoke inhalation
Noise	Deafness

Hazard control for chemicals and **contaminants** may include:

* ventilation and extraction requirements

* respiratory and eye protection

* full body protection

* reducing the length and frequency of exposure.

Recognising hazards

When working on a construction site you need to be aware of other people working on-site and what they are doing. You should know:

* what plant and equipment is being used on-site

* how the work of others could impact on you

* how your work could impact on others.

In order to avoid accidents, you should be aware of the following serious hazards that could occur on-site.

* **Access**: such as missing barriers, tapes and warning signs; obstacles, clutter or spills in walkways.

* **Working at height**: ladders placed too far from work; fragile roof; weight of materials on scaffold too heavy; materials not prevented from falling over the side; guard rails missing.

* **Excavation**: walls of hole or trench not supported – could collapse; no guard rail, tapes or warning signs; heavy materials placed at trench edge.

* **Manual handling**: materials delivered in heavy bags over 25 kilos; no lifting equipment available; walkways cluttered.

* **Plant, machinery and equipment**: machinery guards broken; no training in the use of the equipment; the right equipment for the job not available.

* **Electrical**: underground and overhead cables; cables, leads, plugs or sockets in poor condition; safety inspection of power tools and equipment overdue.

* **Fire**: large quantities of flammable products kept near working area; people smoking; build-up of **combustible** waste on-site; not enough fire extinguishers.

* **Noise**: correct ear protection not available; no warnings given by other workers creating loud noises.

ON FILE

RISK ASSESSMENT

The following table shows extracts from a risk assessment conducted at a small construction worksite. Some of the risk controls suggested involve giving Toolbox Talks. Construction skills produce a set of 70 Toolbox Talks – with a laminated card for each topic.

Hazard	Existing controls	Further controls needed
Working at heights	– Safety induction – PPE supplied and used	Supervision of work at heights
Falling objects	– Safety induction – PPE supplied and used	Regular Toolbox Talks Supervision of work at heights
Fire	– PPE supplied and used – Fire safety training	Nil
Minor accidents (cuts, sprains)	– Trained first aider and first aid kit in each team	Nil
Use of plant and equipment	– Use CITB trained staff to drive vehicles – Plant and equipment inspected and maintained regularly	Nil
Housekeeping	– Induction training – Site inspections – Toolbox Talks	Dust problems: advised use of PPE, increased ventilation, dampening dust before sweeping

TASK

1. Should there always be supervision of work at heights? If so, at what level should the supervisor be qualified? If not, why not?
2. What is included in the Toolbox Talk about falling objects?
3. Would you suggest any further controls for the hazards marked 'nil' in the further controls section?
4. Are the controls suggested for dust control adequate?

NEED TO KNOW...

GO TO...

For information on the web about **risk assessments, safety** and **Toolbox Talks** see the resources on the CD accompanying this book.

HUH?

Risk – the chance of danger

Hazard – danger

Hierarchy – top-down structure, most effective to least effective

Combustible – burns easily

Contaminants – anything that causes pollution

Mastics – tile adhesives

CHECKED

UNIT	Exploring health, safety and welfare in the	BTEC
TWO	construction industry	FIRST
SUBJECT	Hazards and risks	

The main steps of a risk assessment are:

↪ identify the hazards

↪ work out who, how and what?

↪ weigh up the level of risk

↪ select risk control measures

↪ put risk control measures in place

↪ monitor the effect

↪ review regularly

DIFFICULTY RATING	DATE

Factors affecting hazards and risk levels

The risk level associated with a particular hazard changes with different circumstances. For example, the risk of a worker slipping and falling increases when it's wet. Now that you know about some of the hazards to be expected on building sites, you need to consider the *levels* of risk involved and the factors that can increase the risk of an accident happening.

Human factors

Attitude

People who are not committed to their job will not do their best. A careless attitude, or trying to get away with doing the least you possibly can, is not good enough. You need a professional attitude to work in the construction industry – a commitment to doing things right, be positive, respect others, take care and be alert. This doesn't mean you need to be stiff and formal – just mature, focused and positive.

Emotional factors

Every construction worker should have a mature and responsible approach to work and be very safety-conscious and careful at all times. However, nobody is perfect, and there may be occasions when a person's usual attitude is affected by circumstances. Emotional factors, such as anger, **apathy** (couldn't care less), depression, grief, worry and stress can cause a person to lose concentration or motivation. This can in turn affect a person's focus on safety issues, leading to an increased risk of accidents. A good team leader or manager should be alert to the emotional state of team members and where possible alter the work tasks of anyone whose emotional state could interfere with the safety or quality of their work.

Skill level

The skill level of a worker depends on training, learning and the experience of using skills repeatedly in a variety of contexts. It is important that people are assigned to tasks suited to their skills and experience, and that the worker has the confidence to do the tasks safely and well. If you are assigned a task for which you do not have the training, you *must* say so. It would be a serious breach of the health and safety laws for an employer to insist that you do the job in these circumstances. Similarly, if you have received training, but do not yet feel confident or experienced enough to do the job, you should speak up, and ask if you could initially have someone to help or supervise you. Being realistic and honest about your level of ability shows a very mature and responsible attitude to your work and to health and safety.

Responsibility

If you are irresponsible and try to cover up any mistakes you might make, the results could be serious – and the cause could be traced back to you.

You need to take responsibility for yourself, the safety of your work and any mistakes you make. Being honest about mistakes helps to get the problem solved before anything serious happens. You should also keep your actions within the level of your own responsibility – that is according to your job description. Don't try to do things that you are not trained or experienced enough to do.

Experience

You may think that some tasks would be easier or quicker done a different way – but don't be tempted to take short cuts. The methods used in the construction industry have been developed over hundreds of years of trial and error and learning from mistakes. There's always a good reason why experienced people follow all the steps of a task exactly. Gaining experience takes time, but you can speed up the process by practising your skills and by taking the advice of older workers.

Work methods

It takes a while to get used to a new type of tool, machine or piece of equipment. Changing from old equipment, methods and procedures to new ones requires a lot of concentration, and often some re-training, especially in the safe operating procedures for new machines. Workers will not be as quick and efficient with new methods and procedures during the time it takes to get used to the change. If no allowance is made for the extra time required, workers may feel under pressure and try to do more without understanding fully the new procedures or equipment. This can very easily lead to mistakes and accidents.

Skill = Training + Learning + Experience

BRICK TALK

THINK?

The more experience you have of one particular way of doing a job, the harder it is to change. Do you agree?

Do young inexperienced people learn new techniques faster than older people?

How long does it take you to adapt to a new way of doing things and get back 'up to speed'?

Dust is a common workplace hazard. If present, appropriate breathing equipment should be used

→ **Humidity** This is a measure of the moisture content in the air – at high levels of humidity the air feels thick and 'clammy'; some people have difficulty breathing. Humidity is greater in tropical regions. You may have experienced high humidity abroad, in a greenhouse growing tropical plants, or at a heated indoor swimming pool. High humidity contributes to heat exhaustion.

→ **Heat stress** Temperatures are rising in the UK, and a heatwave can cause construction workers to suffer heat stress. The symptoms of this are: muscle cramps, heat rash, severe thirst, fainting; heat exhaustion (**fatigue**, giddiness, **nausea**, headache, moist skin); heatstroke (hot dry skin, confusion, convulsions and eventual loss of consciousness). It is vital to recognise the early signs of heat stress, as heat-stroke can result in death. Report the first signs of heat stress to your supervisor, rest and drink plenty of water.

A S K **The 'no blame' culture accepts that human error is inevitable – we are not perfect – and that procedures have to be changed to reduce the human error that will occur.**

Ask some experienced people in the construction industry what they think about the 'no blame' culture. Does it help to reduce risk?

Workplace factors

Other factors that affect a worker's performance level include weather and environmental conditions. Working in excessive heat, cold or high humidity can affect ability to do the job – for example, if a person's hands are sweaty and slippery, or freezing cold, their grip will be affected. If a person is suffering from heat exhaustion, terrible mistakes could be made. Materials, tools and equipment behave differently at different temperatures.

Visibility may be affected in dusty conditions; air supply and freedom of movement are at risk in confined spaces. Any conditions that are significantly different from normal can cause stress leading to increased risk of hazards causing accidents. To minimise this, workers should be given time to adapt and the correct tools and PPE for the changed circumstances.

Use adequate sun and heat protection to avoid heat stress

ON FILE

ACCIDENT CAUSES

A study of 100 accidents in the construction industry found that worker actions, behaviour or capability contributed to the cause of 70 per cent of the accidents. Workplace factors such as site layout, space and housekeeping contributed to 49 per cent.
Here are four of the accidents.

- A trained worker lost concentration and drilled through his finger with an electric screwdriver – the work area was very small and busy, with people walking by all the time.
- On a very hot day, a worker stopped using gloves while positioning a kerbstone on to a prepared ground – resulting in a crushed finger.
- A worker cut into a gas main believing the gas supply to be cut off, but the main was actually live and a gas explosion was caused.
- A worker used an angle grinder in an enclosed space and the sparks ignited the filter in the extractor fan.

TASK

1. Note down the human factors involved in each of the accidents.
2. Did workplace conditions contribute to any of the accidents? Explain how.
3. What steps would you recommend to avoid these accidents being repeated?

Using gloves can prevent damage or even serious injury to the hands

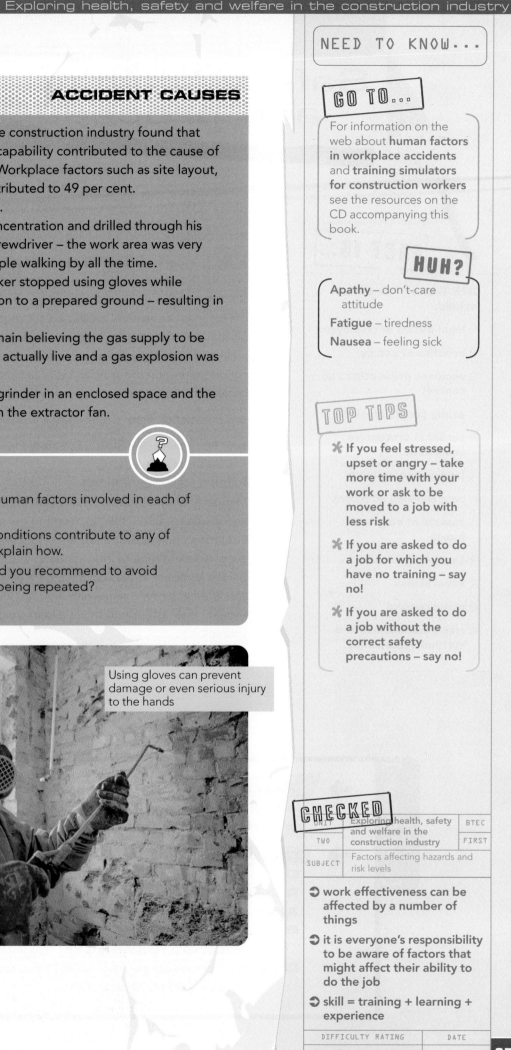

NEED TO KNOW...

GO TO...

For information on the web about **human factors in workplace accidents** and **training simulators for construction workers** see the resources on the CD accompanying this book.

HUH?

- **Apathy** – don't-care attitude
- **Fatigue** – tiredness
- **Nausea** – feeling sick

TOP TIPS

- ✱ If you feel stressed, upset or angry – take more time with your work or ask to be moved to a job with less risk
- ✱ If you are asked to do a job for which you have no training – say no!
- ✱ If you are asked to do a job without the correct safety precautions – say no!

CHECKED

UNIT TWO	Exploring health, safety and welfare in the construction industry	BTEC FIRST
SUBJECT	Factors affecting hazards and risk levels	

- ➲ work effectiveness can be affected by a number of things
- ➲ it is everyone's responsibility to be aware of factors that might affect their ability to do the job
- ➲ skill = training + learning + experience

DIFFICULTY RATING	DATE

Safety training and procedures

Health, safety and welfare training courses and workplace procedures are designed to keep you alive and well. Make sure that you:

* concentrate on your training courses and learn all that you can

* read, understand and follow workplace procedures.

Training

CITB construction skills produce a set of 70 Toolbox Talks. These are 30-minute safety training sessions focusing on a particular type of task or aspect of site safety. Other safety courses may be run at the workplace concentrating on the specific tasks and conditions involved on the project. For example:

* Construction Plant Competence Scheme (CPCS) card

* CITB ConstructionSkills Site Safety Plus courses.

Manual handling

Manual handling training covers safe methods for pushing, pulling, lifting and lowering. Basic manual handling training may be included in **induction** courses, but there are safe manual handling techniques for every different type of job. Handling loads incorrectly can cause severe permanent back injury – so each time you learn a new task, make sure you learn any new manual handling techniques associated with it. You should never attempt to move a load weighing more than 20–25 kg by yourself.

[LOW DOWN]

→ **First aid** This is the basic assistance you can give without medical training to the victim of an accident; first aid aims to keep the person alive until medical aid arrives, prevent the injuries from getting worse and help the person to recover from the injury.

→ **DRABC** – the first aid action plan:

Danger – check for hazards and remove them

Response – is the casualty conscious?

Airway – make sure the casualty's airway is clear

Breathing – is the casualty breathing? If yes, place in recovery position; if no, start expired air resuscitation (EAR)

Circulation – does the casualty have a pulse? If yes, continue EAR; if no, start cardio pulmonary resuscitation (CPR).

When lifting and moving a load, follow these steps for safe manual handling:

1. Make sure you have a clear path to your destination before you pick up the load.

2. Keep your back straight, bend your knees to pick up the load.

3. Use the leg and thigh muscles to straighten the body.

4. Keep your arms close to the body and avoid twisting.

5. Move steadily without rushing.

6. Keep your back straight and bend your knees to place the load down carefully.

Follow the the procedures for safe manual lifting to avoid damaging your back

Work procedures

Work procedures identify the correct steps of a particular task, including:

1. Planning.

2. Selection and checking of tools and equipment.

3. Site preparation.

4. Completing the task.

5. Site clear-up.

6. Cleaning, inspection and storage of tools and equipment.

CHECK IT OUT

Where could you do first aid training?

What items should you find in a first aid box?

Should there be any medication in a first aid box?

ASK

What permits to work are required in your chosen area of the construction industry?

Ask at least three experienced construction workers what preparation may be required before a permit to work would be issued.

Essential PPE

✱ **Head** – hard hat, must fit tightly, report if dropped or hit.

✱ **Feet** – boots, metal toe guard, rubber soles prevent electric shock.

✱ **Ears** – ear defenders – earplugs not good enough for most jobs.

✱ **Eyes** – safety glasses or goggles; welding goggles.

✱ **Face** – safety visor.

✱ **Body** – sturdy overalls, hard to rip, even with sharp edge.

Keep PPE in good condition – keep clean, check 'use-by' dates.

Report faulty equipment – inspect for cracks, dents, tears etc.

Each step will include the recommended safety measures established during a risk assessment, so it is important that you follow the procedures carefully. For example, the recommendations for manually laying a block wall might be:

✱ Use blocks weighing less than 20 kilos.

✱ Ensure blocks are placed as near as possible to the work and on a flat surface.

✱ Use gloves, safety boots and safety helmets.

✱ Ensure the blocks do not have to be lifted above shoulder height.

Permit to work

A permit to work is a form, usually signed by a supervisor or health and safety officer, stating the work to be done and the risk assessment and safety precautions taken. Permits are required for many different types of hazardous work, such as work at heights, work below ground, work in confined spaces and work with flammable materials.

Before issuing a permit for work on a roof, for example, the authorising officer will check:

✱ risk assessments for the particular site and job

✱ fall protection from edges, ladders or scaffold, or through fragile, materials

✱ that equipment for prevention of falls (such as harnesses) is available and in good condition

✱ that workers have received all relevant training

✱ that communication systems are in place (such as two-way radio)

✱ that workers are issued with the appropriate PPE.

Earplugs – for light noise; ear defenders – for loud noise

Dusk mask – for light dust; mask with filter [or respirator] – to protect against fumes

Safety specs [goggles] – protects eyes against dust; visor – protects whole face from dust, chips or sparks

Different levels of PPE are required for different jobs

ON FILE

FREE HEALTH AND SAFETY TRAINING FOR SMALL BUSINESSES

The Health and Safety Executive (HSE) formed a partnership with a local council and a Working Well Together (WWT) group to deliver free health and safety training to small construction companies. Small businesses can rarely afford to lose time from work to attend courses, so the training was run early in the morning. The course covered:

- avoiding slips and trips
- ill health
- working at height
- safe manual handling
- control of substances hazardous to health
- safe site-traffic management
- identifying and working safely with asbestos.

HSE inspectors visited several sites while the training was on offer, conducting inspections and encouraging builders to take advantage of the free training. Many of the builders visited seemed unaware of their legal duties; some who did know about the law ignored it, putting people at risk.

A local builder who attended the course said that health and safety management can be a problem. He said: 'On the course, the hidden costs of having a worker injured were covered, and I see now that good health and safety management could improve my business. I learned how to do a risk assessment – it's much simpler than I thought; and when I've done the CHAS registration (Contractor Health and Safety Assessment Scheme) I think my business will grow.'

TASK

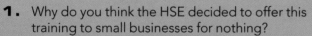

1. Why do you think the HSE decided to offer this training to small businesses for nothing?

2. How could you encourage more small businesses to attend safety training courses?

3. What do you think the 'hidden costs of having a worker injured' are?

4. Why is the builder's business more likely to grow after he has completed the CHAS scheme?

5. Find out more about the 'Working Well Together' organisation. Do you think you will join? Explain why or why not.

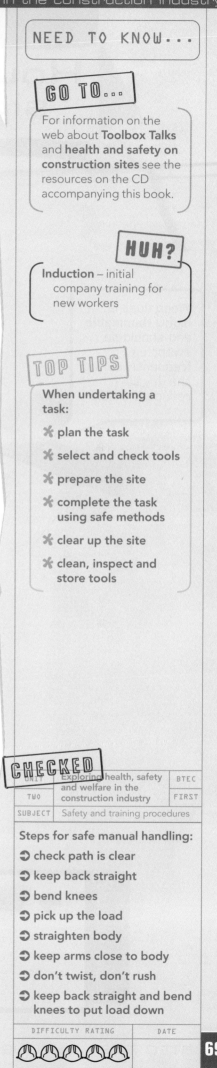

NEED TO KNOW...

GO TO...

For information on the web about **Toolbox Talks** and **health and safety on construction sites** see the resources on the CD accompanying this book.

HUH?

Induction – initial company training for new workers

TOP TIPS

When undertaking a task:

✶ plan the task

✶ select and check tools

✶ prepare the site

✶ complete the task using safe methods

✶ clear up the site

✶ clean, inspect and store tools

CHECKED

UNIT	Exploring health, safety and welfare in the construction industry	BTEC
TWO		FIRST
SUBJECT	Safety and training procedures	

Steps for safe manual handling:

↪ check path is clear

↪ keep back straight

↪ bend knees

↪ pick up the load

↪ straighten body

↪ keep arms close to body

↪ don't twist, don't rush

↪ keep back straight and bend knees to put load down

DIFFICULTY RATING	DATE

Housekeeping, safety signs and fire safety

Housekeeping

Good organisation and tidiness are essential for health, safety and welfare, particularly for prevention of trips, falls and fires. Keeping your workplace clean with plenty of clear space and no clutter makes it much easier to work in efficiently; it looks professional, so encourages professional work and behaviour and, most importantly, it saves lives.

Here are a few basic rules to follow:

* Keep access, exit and emergency routes completely clear at all times.
* Put away tools, equipment and materials as soon as possible.
* Place waste in the correct type of container.
* Recycle as much as possible.
* Dampen down dust before sweeping.

Wood dust is highly **flammable** and should be swept up frequently during the day to reduce fire risk!

BRICK TALK

Safety signs

Employers are required by law to provide safety signs for managing the movement of traffic and pedestrians on-site, and for warning about the presence of any risks. Signs should be placed in an obvious location where they can be clearly seen, away from obstacles, and checked regularly to ensure that they are still in place, still visible and still **legible**. New signs should be explained in induction courses, safety training or site briefings. The different types of sign are explained below.

Mandatory safety signs
Round, with a blue background and white picture or words. You *have to* do what the sign says, such as wear safety goggles.

Warning signs
Yellow triangle, black edging, black picture and writing. The sign warns of a danger to be aware of – such as goods being hoisted overhead.

Prohibitory signs
Red circle, with a line across, white background, black picture and writing. The action on the sign is strictly forbidden as it could increase danger.

[LOW DOWN]

Hazardous waste Some of the waste produced on a construction site is hazardous, and special procedures must be followed for sealing and disposing of the waste product. Hazardous waste that you may encounter in demolition and construction includes: asbestos, fluorescent tubes and paint containing lead.

FIG 2.2: Mandatory, warning and prohibitory signs

Mandatory sign

Eye protection must be worn

Warning sign

Beware of overhead loads

Prohibitory sign

No smoking

Fire safety

A fire will start in any situation where the following are present:

* **Fuel** – this is material that will burn. It could be liquid, gas, dust or solid.

* **Oxygen** – no fire will start or continue burning without oxygen. You can 'suffocate' a fire by removing the source of oxygen – that's why we close doors and windows on a fire.

* **Source of ignition or heat** – this could be a lighted cigarette or a spark from grinding or welding work.

FIG 2.3: The fire triangle – remove one of these three elements, and you will kill the fire

CHECK IT OUT

Talk to your employer or supervisor about how they arrange the work site to give the maximum room for everyone to move around and work. Does your work area arrangement need improving? Include places to store tools, equipment and materials out of the way but close at hand in your plan. Once your area of a work site is set up, keep it as clean and clear as possible.

Type of fire	Correct fire extinguisher
Wood	Red, water
Paper	Red, water
Fabric	Red, water
Petrol	Cream, foam
Oil	Cream, foam
Paint	Cream, foam
Electrical	Black, carbon dioxide
Electrical	Blue, dry powder
Liquids	Blue, dry powder
Gases	Blue, dry powder
Metal	Dry chemical powder such as, salt, graphite and sodium carbonate

All employers should include a fire risk assessment in the workplace health, safety and welfare risk assessment. Fire precautions should include:

* fire detection and warning systems

* emergency escape routes

* provision of adequate firefighting equipment of the correct type

* staff training in fire safety.

Extra fire precautions are needed if the work involves highly flammable and **combustible** products. These 'process fire precautions' may include the provision of fire-protected storage and extraction and ventilation systems.

Firefighting equipment

Firefighting equipment includes extinguishers, buckets of sand or water and fire-resistant blankets. Large buildings may also have automatic sprinklers, hose reels and hydrant systems.

Fire extinguishers

Fire extinguishers are available in four main types. You must make sure that you use the right extinguisher for the type of fire. Never use a water extinguisher on an electrical fire – water conducts electricity and you could receive a fatal shock!

FIG 2.4: Different types of fire extinguisher

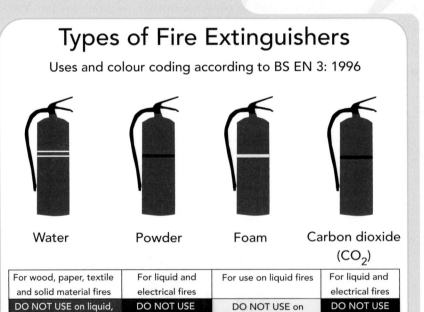

Types of Fire Extinguishers

Uses and colour coding according to BS EN 3: 1996

Water	Powder	Foam	Carbon dioxide (CO$_2$)
For wood, paper, textile and solid material fires	For liquid and electrical fires	For use on liquid fires	For liquid and electrical fires
DO NOT USE on liquid, electrical or metal fires	DO NOT USE on metal fires	DO NOT USE on electrical or metal fires	DO NOT USE on metal fires

The contents of an extinguisher are indicated by a zone of colour on the red body of the extinguisher.

ON FILE

GAS CYLINDERS

There were 44 fire incidents involving **acetylene** gas in London between May 2003 and May 2005. One fire on a construction site in Kings Cross involved two cylinders of acetylene gas. Thousands of residents and workers were evacuated, and rail services were closed for two days. A new campaign led by the London Fire Service is asking for stricter controls on the use of acetylene gas cylinders, including signage and storage. Heating or mechanical shock can cause internal heating of acetylene gas within the cylinder, and the risk of explosion lasts for 24 hours.

According to the Fire Safety Advice Centre, there are more than ten fires on construction sites every day. To avoid fires the following advice should be followed:

- Turn all gas cylinders off after use, check and maintain cylinder fittings regularly.
- Store all gas cylinders outside buildings in well-ventilated and secure areas.
- Store solvents, adhesives and other flammable materials in lockable steel storage containers.
- If a leak is suspected, stop using the gas – and report the leak.
- Follow procedures, such as permits-to-work for all hot work (welding); this should include having extinguishers close by – also check that sparks have not landed on anything combustible.
- Keep the site tidy; clear away rubbish promptly.
- Keep only a one-day supply of any highly flammable materials on-site.
- Do not burn waste materials on site.
- Strongly enforce 'no smoking' rules.

TASK

1. What types of gas are used in construction work? Find out the storage recommendations for each type of gas.

2. Read through the recommendations for fire safety given by the Fire Safety Advice Centre; what further safety measures do you think are necessary to prevent gas fires and explosions?

3. Make a list of all the flammable and combustible materials you might find on a construction site. How should they be stored? What should you do with any unused or waste amounts of these materials?

4. How can you make sure that workers follow the no smoking rules on site? What penalties should apply to workers found smoking near flammable or combustible materials?

NEED TO KNOW...

GO TO...

For information on the web about **safety signs, fires, explosions** and **hazardous substances** see the resources on the CD accompanying this book.

HUH?

Legible – clearly readable

Acetylene – a highly flammable gas used in welding

Combustible – capable of burning

Flammable – easily set on fire

TOP TIPS

✳ Know your workplace fire procedures and how to use fire-fighting equipment

✳ Tidiness saves lives: keep all walkways clear, put everything away in the right place (including waste), dampen dust before sweeping

CHECKED

UNIT TWO	Exploring health, safety and welfare in the construction industry	BTEC FIRST
SUBJECT	Housekeeping, safety signs and fire safety	

➲ Mandatory signs: must be done – blue and white

➲ Warning signs: danger notice – yellow and black

➲ Prohibitory signs: must not be done – red, white and black

DIFFICULTY RATING	DATE

Being safe in hazardous places

In the construction industry you may be required to work at height, below ground and sometimes in confined spaces. Each of these types of location is more hazardous than working at ground level or in relatively open spaces, and special safety procedures must be followed.

Working at height

Falls from height cause half of all the fatal accidents on construction sites. The Work at Height Regulations 2005 apply to all types of work at height where there is a risk of a fall that could cause personal injury. As an employee working under supervision, you are required by law (Work at Height Regulation 14) to:

* report safety hazards to your employer or supervisor

* use all the equipment and safety devices properly and follow the procedures demonstrated in your training or instructions

* consult your employer or supervisor if you think the equipment or the procedures are unsafe.

Risk control measures for avoiding falls from height include:

* using work platforms and scaffolds installed or erected by competent persons

* making sure there is adequate edge protection, guard rails and toe boards on work platforms and scaffolds

* using mobile elevating work platforms for fragile roofs, where possible; being sure that the plant operator is trained and competent

* using harnesses and safety nets on work platforms on fragile roofs

* keeping materials close at hand

[LOW DOWN]

→ **Breathing apparatus** Breathing apparatus is used to supply clean air to workers in confined spaces, or in dust-, smoke- or fume-filled areas. Air-line apparatus delivers compressed air from a cylinder to a mask fitted over the worker's face. Some air-line apparatus has lines running to the work area from cylinders outside – the cylinders must be checked and changed regularly. There are also self-contained breathing apparatus packs, where a small cylinder is strapped to the worker's back. Special training is required before a person can safely use air-line breathing apparatus.

→ **Isolation** Removing the source of electrical, gas, water or fuel supply so that equipment, machinery, pipework or cabling is completely inactive. Electrical isolation means removing the power source; isolating services means cutting off, for example, the gas supply, so that work can be done on the pipe.

* lowering and raising materials and waste using safe hoisting systems

* never working in wet, windy or other extreme weather conditions

* making sure there is adequate edge protection on sloping roofs

* only using ladders for short tasks involving no heavy work.

When using ladders, check that the ladder is in good condition before securing it at a 75-degree angle, with at least two rungs above the resting height. Never over-reach from a ladder – get down and move the ladder closer to the work.

Working below ground

Trenches are excavated on construction sites for foundations, pipelines and other utility services. If the trench is not excavated and supported according to a safe-work plan, there is a danger that the sides could collapse, burying the workers below ground, or that the excavation could cause surrounding structures to collapse. Other dangers with excavating trenches include:

* breaking electric cables, gas pipelines or other underground services while excavating

* materials falling into the trench (stacked too close to the edge)

* people or vehicles falling into an unguarded or unsignposted trench.

To ensure safe work on excavations:

* the excavation should be planned by a suitably qualified person

* all excavation work should be supervised by a competent person

* the location must be checked for underground utilities, especially electricity

* warning barriers and signs should be placed around the excavation area – to keep other workers, the public and vehicles clear of the site

* stop-blocks must be placed at the correct distance to stop any plant working on the excavation from coming too near to the edge

* plant and vehicle exhaust fumes must not reach the excavation

* sides of the excavation should be dug at a slight angle and/or a secure lining of timber boards placed all around the internal walls

* toe boards and guard rails should be secured in place around the edge of the excavation

* safe access to the base of the excavation must be installed, such as a secured sloping ladder

* everyone in the excavation must wear a hard hat.

CHECK IT OUT

Is there any work at height involved in your chosen area of the construction industry?

Research the different types of breathing apparatus available.

What training is required before you are qualified to use the apparatus?

Which construction industry workers use breathing apparatus regularly in their work?

ASK Talk to experienced painters and decorators and find out the answers to the following questions.

What hazards are involved when using paint in an unventilated room?

Which types of paint or other materials used by painters produce hazardous fumes?

What safety precautions do painters apply to avoid being overcome by fumes?

What should you do if a colleague collapses in a fume-filled room?

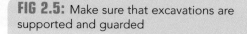

FIG 2.5: Make sure that excavations are supported and guarded

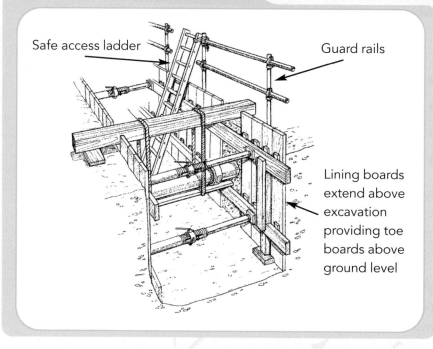

Safe access ladder

Guard rails

Lining boards extend above excavation providing toe boards above ground level

Working in confined spaces

A confined space is one that is enclosed, such as open-topped chambers (manholes), ducts and unventilated rooms. In these spaces there is little or no ventilation, and a worker is in danger of being overcome by fumes from welding or painting, the lack of oxygen, a build-up of dust or increasing heat. A risk assessment for work in confined spaces should include:

* the work environment

* the materials and tools to be used

* the training, experience and health of the workers

* emergency rescue procedures.

The Confined Spaces Regulations 1997 specify three key duties towards those working in confined spaces.

1. To arrange, wherever possible, to complete work from outside avoiding entry into any confined space. If this cannot be arranged:

2. To follow a safe system of work.

3. To have emergency arrangements in place before starting work.

Safe system of work

Safe work systems include the following precautions

* A supervisor must be present.

* Check the training, confidence, health and size of workers (small-framed workers will have more manoeuvrability).

* Isolate mechanical and electrical equipment and pipework.

* Ensure adequate entrance size – easy access in an emergency, clean space.

* Prepare ventilation and breathing equipment (air-line).

* Have the air tested by a qualified person.

* Select tools and equipment carefully. Never, ever, use petrol-fuelled equipment – these produce deadly **carbon monoxide**. Use non-sparking tools, extra-low-voltage equipment (less than 25 volts) and residual current devices.

* Prepare emergency arrangements: provide equipment such as harnesses, lifelines, two-way radios and an agreed alarm signal; arrange rescue practices before starting work.

ON FILE

DANGEROUS EXCAVATION

A bookshop with residential accommodation above collapsed after building work started on the vacant site next door. Excavations for the foundations on the vacant site involved digging out a trench parallel to the bookshop wall. Neither the excavation nor the shop wall were supported. During the night, one resident noticed cracks appearing on the wall of his flat – all the occupants were evacuated and the building later collapsed on to the empty site. The company was fined £90,000, the director was also personally fined £90,000 plus £3,000 compensation awarded to each of the occupants of the building.

TASK

1. Find out what structural support for the wall of the adjacent building should have been installed before excavations began.

2. What supports should have been present inside and around the excavated trench for the foundations?

3. What could have happened if the building had collapsed when people were working inside the foundation trench?

4. Do you think the penalty is fair? Explain your answer.

NEED TO KNOW...

GO TO...

For information on the web about **working at height, below ground or in confined spaces** see the resources on the CD accompanying this book.

HUH?

Carbon monoxide – a colourless, odourless gas that can kill

Shoring – support for a structure or excavation wall

TOP TIPS

✳ When working at height, in an excavation or in a confined space, make you know, and follow, all the safety precautions

✳ If you have any worries, report them at once

CHECKED

UNIT	Exploring health, safety and welfare in the construction industry	BTEC
TWO		FIRST
SUBJECT	Being safe in hazardous places	

Regulations applying to dangerous working environments include:

➲ Working at Height Regulations 2005

➲ Confined Spaces Regulations 1997

DIFFICULTY RATING	DATE

Being safe with plant, equipment, machinery and electricity

PUWER

The Provision and Use of Work Equipment Regulations (PUWER) 1998 outline the legal requirements for preventing and controlling risks from tools, equipment and plant used at work. This includes hand tools, electronic equipment, ladders, vehicles, power tools and scaffolds. The employer must ensure that tools, equipment and plant either provided by them, or brought into work by the employee, is:

* well-maintained, inspected regularly by a competent person and safe to use

* used in conjunction with recommended safety measures, such as guards, appropriate PPE and warning signs

* used only by suitably trained people.

The use of tools, equipment and plant includes activities such as: cleaning, operating, maintaining, repairing and starting or stopping the equipment. The PUWER regulations prevent young people under 18 from operating equipment such as power presses or hand-fed timber sawing and planing machinery 'unless they have the necessary maturity and competence'.

Electricity

Managing the risks associated with electricity on a construction site requires very careful risk assessment, analysis and planning. Outdoor work and wet conditions increase the risk of electric shocks and the degree of injury. Electricity supply has to be moved as work progresses, introducing more possibilities for error. Excavations and demolition can cause damage to on-site electrical supply or previously installed services, such as underground or overhead cables. Trailing cables can also be a hazard, and may be damaged by vehicles. Safe work methods to avoid electric shock on construction sites include the following:

* Make sure that the electrical equipment is safe – there should be an inspection tag or record stating when the equipment was last checked.

* Use battery-powered tools or 110 volt supply whenever possible.

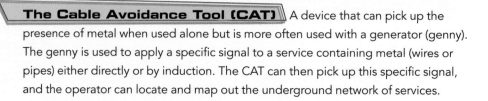

NEED TO KNOW...

GO TO...

For information on the web about **electrical safety** see the resources on the CD accompanying this book.

✱ Make sure that excavation is planned by a qualified person, and that the locations of underground cables have been identified. The site should be checked using a **Cable Avoidance Tool**, or CAT, a device for locating cables. Be aware that cable may have been laid by an independent contractor and there might be no record of the location.

✱ Be aware of the location of overhead cables, particularly when working at heights or using plant such as cranes or mobile elevating work platforms.

TOP TIPS

To ensure electrical safety:

✱ Check electrical equipment is safe (look for tag and date)

✱ Use battery-powered tools or use 110 volt supply

✱ Check excavations for underground cables

✱ Check for overhead cables

ON FILE

OVERHEAD POWER LINES

Safe working with electricity could avert some of the fatal accidents on construction sites. The Working Well Together website lists the following accidents where workers received electric shocks from contacts with overhead power lines.

- A painter was moving a tower scaffold, which was beneath the overhead lines at the time.
- A plant operator was erecting a mobile lighting gantry on road surfacing work, and the gantry came into contact with overhead power lines.
- A scaffolder made contact with overhead power lines during erection of a scaffold.
- A plant operator was lifting a portable accommodation unit using a lorry mounted crane, which made contact with overhead power lines.

TASK

1. Which of the accidents listed above would have been fatal? Why?

2. Consider each accident and discuss with your colleagues and supervisor or trainer the steps that could have been taken to avoid each one.

3. What are the electrical hazards in your chosen area of the construction industry?

4. Find out the safe work method for one of the electrical hazards in your chosen construction area.

CHECK IT OUT

What tools and equipment do you use in your chosen area of the construction industry?

List the tools and equipment and the regularity of PUWER checks for each one.

Choose three tools or pieces of equipment – what are the key points checked when each one is inspected?

CHECKED

UNIT	Exploring health, safety and welfare in the construction industry	BTEC
TWO		FIRST
SUBJECT	Being safe with plant, equipment, machinery and electricity	

➔ PUWER stands for Provision and Use of Work Equipment Regulations

➔ PUWER 1998 prevents and controls risks from equipment, plant and tools used at work

DIFFICULTY RATING	DATE

Unit summary

✔ Avoid accidents!

Accidents are caused by unsafe conditions or unsafe acts. The most frequent accidents in the construction industry are:

* falls from height (through a fragile roof or skylight; from a ladder or a scaffold)

* being hit with force (by moving vehicles, falling loads or equipment, collapsing structures).

✔ The law protects you!

Under the Health and Safety at Work etc. Act 1974 (HASAWA) both the employer and the employee have responsibilities. Your responsibilities include being aware of hazards and following procedures, such as using the correct PPE and safe work methods. Employers should 'take all steps, as far as reasonably practical, to ensure your health, safety and welfare at work'. Penalties for offences averaged £9,615 in 2005/2006.

✔ Hazards and risks

A hazard is a danger – an accident waiting to happen. Level of risk is the likelihood of an accident happening and the degree of injury that could be caused. The Management of Health and Safety at Work Regulations 1999 require employers to assess risks and specify control measures to prevent accidents. You need to recognise hazards at work, such as those associated with: access; working at heights; materials; excavation; manual handling; plant, machinery and equipment; electricity; fire and noise.

✔ Factors affecting hazards and risk levels

The likelihood of an accident may change under different conditions; factors affecting risk levels include: work methods; new tools, equipment, materials or machinery; new procedures; workplace factors, such as temperature, dust, humidity and confined spaces; human factors such as attitude, training and learning, responsibility and experience.

✔ Safety training and procedures

These are designed to keep you alive and well. Training may include induction, site safety courses and Toolbox Talks. Manual handling training is important for all types of work – handling loads incorrectly can cause severe permanent back injury. Work procedures specify the steps of a job: planning; selecting and checking tools and equipment; site preparation; completing the task; site clear-up; and cleaning, inspection and storage of tools and equipment. Some work will require a permit to work, which specifies the work to be done, the person authorised and trained to do the work, and the safety precautions taken.

Housekeeping, safety signs and fire safety

Good organisation and tidiness make a workplace safe and easier to work in. They encourage professional work and behaviour, and save lives. There are three main types of safety sign that should be used to prevent, or warn of, hazards: these are mandatory, warning and prohibitory. Fire safety precautions should include fire detection and warning systems, emergency escape routes and provision of adequate fire extinguishers.

Being safe in hazardous places

There are special precautions and procedures required for: working at heights (such as using safe scaffolds, elevated work platforms, safety nets and harnesses); working below ground (such as shoring sides of trenches and installing barriers, toe boards and stop blocks); and working in confined spaces (such as ensuring adequate entrance size, preparing ventilation and breathing equipment, testing air quality, preparing emergency rescue arrangements).

Being safe with plant, equipment, machinery and electricity

The Provision and Use of Work Equipment Regulations (PUWER) 1998 require that tools, equipment and plant used at work must be: well-maintained; inspected regularly and safe to use; used with safety measures such as guards, PPE and warning signs; used by trained people. Some safe work methods to avoid electric shock on construction sites include: using only safe electrical equipment; using battery-powered tools or 110 volt supply; checking the location of underground cables; being aware of overhead cables.

When you have completed this unit, you should be able to:

✔ identify and describe the effects of gravitational forces on construction materials including using supporting calculations

✔ identify and describe the effects of temperature change on construction materials including using supporting calculations

✔ identify and describe practical methods used to accommodate the effects of temperature changes

✔ describe beneficial practical applications of temperature effect on materials

✔ solve practical construction problems by transposing formulae and evaluating formulae numerically

✔ solve practical construction problems using trigonometry and graphical methods.

The use of science and mathematics in construction

Scale: 1:1

Coming up in this unit...

Science and mathematics underpin many of the activities undertaken by professionals in the construction industry. Many jobs now demand more than just the ability to perform practical tasks, important though these skills are. Imagine, for example, having to make and fit a window frame. No matter how good your hands-on joinery skills, if you cannot manage the measuring and calculation needed to ensure you cut the right amount and size of wood, and cut the correct angles, you won't be able to complete the job satisfactorily. In your construction career you will probably be called upon to perform mathematical calculations to do with dimensions, forces, areas, volumes, material quantities and costs.

For many jobs in construction you will have to go further than simply understanding basic measuring and calculation. In order to choose the right materials for a particular job, for instance, you need to understand how different materials respond to forces such as stress and strain, and environmental factors such as moisture and temperature.

The aim of this unit is to introduce the basic mathematical techniques that you will use on a daily basis as a construction professional. It will also give you a sound foundation in the basic scientific concepts used in the construction industry.

Maths is used in all aspects of the construction industry

The materials you use in construction have different characteristics, such as heaviness, strength, flexibility, **elasticity**, **brittleness**, **combustibility**, **porosity** and heat conduction. The characteristics of a material depend on its molecular content and structure and its mass, density and volume. Because different materials vary so much in structure and characteristics, they respond differently to **forces** or **loads**. A force or load may be the pressure (compression force) of the weight above the material, such as the weight of a multi-storey building pressing on its foundations; pulling or stretching forces, such as the tension on the cables of a suspension bridge; or pushing forces, such as strong wind.

The mass, volume and density of materials

In order to choose the right material for a job you need to understand the way different materials react to forces and loads. The basic characteristics of a material involve its mass, volume and density.

[CONCEPTS]

➤ **Mass** is the amount of matter (solid particles) in a compound; measured in kilograms (kg).

➤ **Volume** is the amount of space the material occupies; measured in cubic metres (m^3).

➤ **Density** is the amount of mass per unit volume of a material, measured in kilograms per cubic metre (kg/m^3).

Mass

Mass is different from weight – the same object will weigh a different amount on Earth than on the Moon because weight is affected by **gravity**, mass is not, and the pull of gravity is greater on the Earth than on the Moon.

* On Earth a bag of cement has a mass of 50.8 kilos and weighs 50.8 kilos.

* In space a bag of cement has a mass of 50.8 kilos and weighs nothing!

Volume

The volume of a solid object is the three-dimensional idea of how much space it takes up. This can be expressed as a number in units cubed (e.g. m^3). You can calculate the volume of straight-edged and circular shapes using simple arithmetic formulae – we'll look at these later. You might want to distinguish between volume and capacity, for instance if you need to know how much water your bucket will hold when you are mixing cement. Capacity describes how much liquid the container can hold (usually measured in litres), and volume describes how much space it takes up (usually measured in cubic metres).

Density

Density equals mass divided by volume. A one kilo bag of roof insulation is much larger in volume than a one kilo bag of nails. The matter in the roof insulation is less tightly packed – it has fewer grams of matter per cubic centimetre (g/cm^3).

Forces and loads acting on materials

A force is anything that makes an object move and includes gravity, pressure and tension. A load is the overall force a structure is subjected to, including supporting weight, mass and outside forces, such as pressure from wind. Every force or load has an equal and opposite reaction – the resistance you feel when you push an object is the force of the object pushing back at you.

Gravity

Why do astronauts float about in space, weightless, and yet when they are back on Earth, they are firmly planted on the ground? The answer is gravity – the pulling force of a larger object (such as the Earth) on a smaller object (such as you and me). If you pick an object up, you feel this pulling force of the Earth on the object as weight.

Pressure

Pressure can be positive or negative. Positive pressure includes the pushing force of wind, or a bulldozer, and compressive forces where objects are being squashed or pushed down. The columns, walls or foundations holding up a building are in compression. Negative pressure acts as suction.

Tension

Tensile forces act where objects are stretched, for example the wires holding up a suspension bridge, or the ropes hauling up a load.

Tensile forces are utilised in the design of suspension bridges

TRY THIS

Density = mass ÷ volume.

A typical brick has a mass of 2,268g and occupies a volume of 1,230cm³.

What is the density of the brick?

FACT IS...

The densities of air and some common materials in g/cm³:

air	0.0013
oak wood	0.6–0.9
water	1.00
brick	1.84
aluminium	2.70
steel	7.80

Loads on a building

The loads that exert forces on structures built include:

* dead loads – the total weight of all parts of a structure which are permanent and stationary, such as walls, beams, columns and floor slabs

* imposed or live loads – additional loads which act on a structure during its use, such as people, furniture, machinery and cars

* snow load – snow adds a considerable load, particularly on roof structures – you need to think about where the snow will settle and the shape and the pitch of the roof

* wind load – wind can produce positive pressure (pushing), negative pressure (suction) and uplift forces on roofs – lightweight buildings are at the most risk of wind load.

Loads can be:

* distributed – the load is spread over the full area or length

* concentrated or point – the loads are concentrated at one point.

WATCH IT!

This is what could happen if you miscalculate loads and forces:

* crushing
* buckling
* settlement
* sliding
* overturning
* bending
* fatigue
* creep.

GROUND RULES: FORCE

Force is measured in Newtons (N) using the equation:

force = mass x acceleration (F = ma)

Gravitational force is calculated using the equation:

force = mass x (a)g

where (a) is the acceleration due to gravity and is equal to 9.8 metres per second squared ($9.8 m/s^2$).

HOW IT WORKS...

The force of gravity acting on a 1kg object is 1 x 9.8 = 9.8N.

If a bucket of sand hanging from a rope has a mass of 30kg, then the force stretching the rope (or the *tension* in the rope) is 30 x 9.8N.

NOW TRY THIS...

Question 1

A bucket of cement of mass 40kg is tied to the end of a rope connected to a hoist. Calculate the tension in the rope when the bucket is suspended but stationary. Take the gravitational field (g) as 9.81N/kg.

FIG 3.1: Pulley system

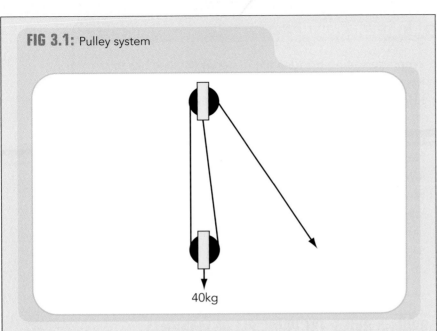

40kg

In Figure 3.1 the tension in the rope is the same as the force acting on the rope.

The force acting vertically downwards due to the weight of the bucket must be equal to the force acting upwards on the rope, i.e. the tension, T.

Weight of the bucket of cement, F = mg

Question 2

The tension in a rope lifting a crate vertically upwards is 2800N. Determine its acceleration if the mass of the crate is 270kg.
Remember F = ma.

[CONCEPTS]

Hooke's law of elasticity describes how the strain on an elastic material is related to the stress placed on the material.

Stress is the force (per unit area) exerted on the internal structure of a material.

Strain is the way in which the size or shape of a material is changed by stress, for example, when a spring is pulled it gets longer. Normal strain involves changes in dimension but not shape; shear strain always involves changes in shape, and can include changes in dimension.

Modulus of elasticity λ is how much elasticity an object or substance has; this tells us how likely a material is to be deformed when a force is applied to it and whether or not it will return to the original shape when the force is removed. The elastic modulus of an object is defined as the slope of its stress-strain curve.

Factor of safety (FoS) is an amount factored in to ensure the safety and stability of a structure when withstanding loads.

When designing bridges the material you use has to have a safe margin of error built in to ensure that the bridge never fails when in use. If the material you choose has to withstand a load of 200 Newtons, you might decide to make sure it can actually withstand 800 Newtons, just to be on the safe side. The factor of safety is the strength of the material, divided by the load it has to withstand and usually ranges from 1.25 to 4. The lower factor of safety is used when materials and conditions are known in detail. Higher factors of safety are used when there is less certainty. In the example above, the factor of safety is 800/200 = 4.

GROUND RULES: THE PRINCIPLE OF MOMENTS

A moment of force is the turning power of that force. Examples of a turning force include opening a door, using a spanner and turning a steering wheel. The simplest example to explain the principle of moments is a **pivot** such as that found on the digging arm of an excavator. Forces can make objects turn if there is a pivot. Think of a playground see-saw – effectively a simple plank with a pivot at its centre. When no one is on it the see-saw is level, but it tips up if someone gets onto one end. It is possible to balance the see-saw again if someone else gets onto the other end. Even if they are a different weight as long as they sit in the right place, they can balance the see-saw again. This is because of the principle of moments.

Imagine that a force of 10N acted on a see-saw 2m from the pivot. This is how we would work out the moment:
moment = force x distance = 10 x 2 = 20Nm

HOW IT WORKS...

A see-saw will balance if the moments on each side of the pivot are equal. This is why you might have to adjust your position on a see-saw if you are a different weight from the person on the other end.

If the beam on the pivot is in equilibrium (level) then the sum of the clockwise moments of force is equal to the sum of the anti-clockwise moments of force. The forces pushing one way are the same as the forces pushing the other way.

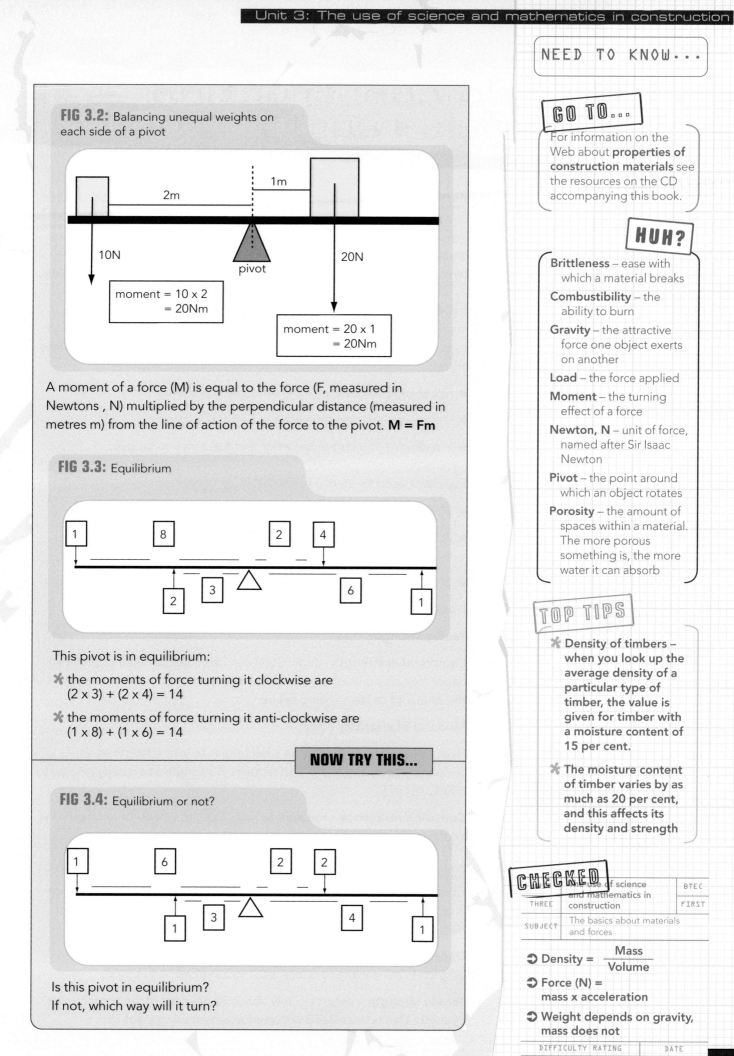

FIG 3.2: Balancing unequal weights on each side of a pivot

1m

2m

10N

pivot

20N

moment = 10 x 2
= 20Nm

moment = 20 x 1
= 20Nm

A moment of a force (M) is equal to the force (F, measured in Newtons , N) multiplied by the perpendicular distance (measured in metres m) from the line of action of the force to the pivot. **M = Fm**

FIG 3.3: Equilibrium

1   8   2   4

2   3   6

1

This pivot is in equilibrium:

* the moments of force turning it clockwise are
  (2 x 3) + (2 x 4) = 14

* the moments of force turning it anti-clockwise are
  (1 x 8) + (1 x 6) = 14

NOW TRY THIS...

FIG 3.4: Equilibrium or not?

1   6   2   2

1   3   4

1

Is this pivot in equilibrium?
If not, which way will it turn?

NEED TO KNOW...

GO TO...

For information on the Web about **properties of construction materials** see the resources on the CD accompanying this book.

HUH?

Brittleness – ease with which a material breaks

Combustibility – the ability to burn

Gravity – the attractive force one object exerts on another

Load – the force applied

Moment – the turning effect of a force

Newton, N – unit of force, named after Sir Isaac Newton

Pivot – the point around which an object rotates

Porosity – the amount of spaces within a material. The more porous something is, the more water it can absorb

TOP TIPS

* Density of timbers – when you look up the average density of a particular type of timber, the value is given for timber with a moisture content of 15 per cent.

* The moisture content of timber varies by as much as 20 per cent, and this affects its density and strength

CHECKED

	The use of science and mathematics in construction	BTEC
THREE		FIRST
SUBJECT	The basics about materials and forces	

➲ Density = $\dfrac{\text{Mass}}{\text{Volume}}$

➲ Force (N) = mass x acceleration

➲ Weight depends on gravity, mass does not

DIFFICULTY RATING   DATE

How do construction materials react to force?

You need to choose the most cost-effective material for each job. Builders cannot afford to use the strongest material for every job, for example, high-strength steel may be an essential framework for structures bearing great loads, but for a single storey building, a timber frame may be the best. Here are some common construction materials and their properties.

Steel

Steel is made from iron with between 0.2 and 1.7 per cent carbon. High carbon content makes steel harder but more brittle. Other additions may include:

* nickel and manganese – to increase tensile strength

* chromium – to increase hardness and melting point

* vanadium – to increase hardness and resistance to fatigue.

Standards used for steel are:

* 2005 AISC Specification for Structural Steel Buildings

* 2004 RCSC Specification for Structural Joints Using ASTM A325 or A490 Bolts

* 2005 AISC Code of Standard Practice for Steel Buildings and Bridges.

Density – high, 7.8g/cm^3

Tensile strength – high.

Compressive strength – high. Steel can be subject to shearing forces under high compression loads. Shearing forces act parallel to the plane of the steel and cause a sliding failure.

Modulus of elasticity – high.

Heat tolerance – high. Carbon steel begins to lose strength at temperatures above 300°C and reduces in strength at a steady rate up to 800°C.

Corrosion resistance – medium to high. Stainless steel containing nickel and chromium is highly resistant to corrosion and fire.

Concrete

Concrete is a low-cost material manufactured using cement, aggregate and water. Used for foundations in buildings and reinforced with steel bars and mesh for use in beams.

Density – varies. Depends on the mix. High density concrete is about 2.5g/cm^3; low density concrete is 0.4 to 2.0g/cm^3.

Tensile strength – very low. Only about 10 per cent of the compressive strength. Can be increased with reinforcement using steel.

Tensile strength is usually measured in thousands of pounds per square inch (KSI)

BRICK TALK

Compressive strength – varies. Depends on the type of aggregate used, air content and the free water to cement ratio. Reducing the free water to cement ratio increases the strength of concrete. If the mass of water in a concrete mixture is 40kg and the mass of cement is 80kg. The free water to cement ratio is 40/80 = 0.5.

Modulus of elasticity – high. Concrete is very stiff and brittle.

Heat tolerance – high.

Corrosion resistance – relatively low. Concrete contains calcium carbonate, which is soluble in acids, such as rainwater. It also reacts with iron oxide (rust).

Brick

Bricks are durable, low-maintenance and attractive building materials made from clay, sometimes with other chemicals added during manufacture. The clay is pressed into moulds and fired at temperatures up to 1000°C. Most bricks are rectangular; the most common size is 215 x 102 x 65mm.

Density – varies. Generally around 1.85g/cm^3

Tensile strength – low.

Compressive strength – ranges from 4 to 180N/mm^2. Water content reduces compressive strength and thermal resistance.

Modulus of elasticity – very high.

Heat tolerance – very high.

Corrosion resistance – medium. Bricks can be subject to frost and salt damage.

Plastics

Produced as a by-product of the oil industry, plastics are very lightweight and do not absorb water, therefore they are not affected by frost. Plastics are used for used for: pipes, damp-proof courses, window frames, floor coverings, fillers and sealants, plugs, sockets, thermal insulators (plastic foam), light fittings and cable insulation. They are used because they are good electrical insulators. Unplasticised polyvinylchloride (uPVC) is resistant to chemicals and can be used for underground pipes.

There are two main types of plastic:

* thermoplastic – becomes soft when heated and hardens on cooling

* thermosetting – does not soften when heated but can char with excessive heat.

Density – varies from 0.9 to over 2.0g/cm^3. (The density of water is 1g/cm^3 or 1kg/m^3)

Tensile strength – low.

Compressive strength – low. Plastics cannot be used to bear loads.

Modulus of elasticity – low.

Heat tolerance – low. Plastics tend to melt or char at relatively low temperatures. They also have high levels of thermal movement (high coefficient of linear expansion).

Corrosion resistance – high. Plastics are generally resistant to corrosion, except when used externally where some types can be degraded by ultraviolet (UV) light.

Aluminium alloys

Aluminium is a light metal with a low tensile strength that is resistant to corrosion. It is used for gutters and flashings. Alloys of aluminium may contain copper, zinc, manganese, silicon, or magnesium. These additions, and treatment such as **tempering**, increase its mechanical properties. Aluminium alloys are used for lightweight frames, roofing and cladding.

Density – relatively low. 2.7g/cm³

Tensile strength – medium to high. It is less than steel, but greatly improved by alloying.

Compressive strength – high.

Modulus of elasticity – high.

Heat tolerance – medium. Not as high as steel, due to its lower melting point, but still high enough for most construction purposes. It expands on heating, so this needs to be taken into account. It is a good heat and electricity conductor.

Corrosion resistance – high. Aluminium develops a thin but tough natural protective oxide coating. Surface treatment such as anodising, painting or lacquering can increase its resistance. It will corrode in strong acid.

Glass

Glass is made from quartz sand, which is almost 100 per cent crystalline silica (silicon dioxide, SiO_2). Most types of glass contain about 70 per cent silica. Soda-lime glass contains almost 30 per cent sodium and calcium oxides or carbonates. Glass is used for glazing windows and doors, glass fibre is used in optical cables, to reinforce plastic products and in thermal insulation. The properties of glass can be modified or changed with the addition of other compounds or heat treatment.

Density – varies. From less dense than aluminium to more dense than iron. Average 2.5g/cm³

Tensile strength – low. Untreated glass is very brittle and breaks into sharp shards under tension.

Compressive strength – high. pure glass can withstand high compressive forces.

Modulus of elasticity – high.

Heat tolerance – high. Glass has a very high melting point.

Corrosion resistance – very high.

Timber

Timber is a renewable construction material used for framework, joists, studs, floorboards and doors. There are two types of timber:

* softwood – from coniferous trees, such as pine

* hardwood – from deciduous trees, such as ash and oak.

The characteristics of timber vary greatly.

Density – relatively low and varies depending on type. Usually between 0.5 and 0.7g/cm³. The higher the density, the more strength the timber has.

Tensile strength – high strength to mass ratio.

Compressive strength – high strength to mass ratio. The dryer (more seasoned) the wood, the higher the strength. The strength of timber with a moisture content of 30 per cent is only about two-thirds the strength at 12 per cent moisture content.

Heat tolerance – low compared to other construction materials. But thick timber has relatively good structural integrity in fire.

Corrosion resistance – low. Timber has low resistance to damp, fungal and insect attack. Wet and dry rot causes decay and loss of strength.

Other materials

The following are also commonly used in construction:

* copper – used for roofing, flashing, pipes

* iron – used for boilers, radiators

* lead – used for roofing, flashing

* zinc – used for roofing, flashing and to protect steel.

WATCH IT!

Choosing the wrong material can result in failure. Types of material failure include:

* **overload – material fails or breaks when put under undue strain, such as heavy snowfall**

* **fatigue – caused by exposure to stress over time, weakening the material to breaking point**

* **creep – very slow change in shape or deformation in response to stress.**

NEED TO KNOW...

GO TO...

For information on the Web about **forces**, **loads** and **moments** see the resources on the CD accompanying this book.

HUH?

Anodising – the electrical coating of aluminium with oxide

Degradation – the decay of a material

Tempering – the heat treatment of metal to increase hardness or strength

TOP TIPS

* Clay bricks are usually hollow, better for thermal resistance

* Engineering bricks have low porosity and high density, good for use where resistance to degradation and high compressive strength is demanded

* Materials vary, check the specifications to find out a material's exact mechanical characteristics

CHECKED

UNIT	The use of science and mathematics in construction	BTEC
THREE		FIRST
SUBJECT	How do construction materials react to force	

→ High modulus of elasticity means a material is stiff and inflexible – the material may be brittle and break easily

→ High density materials are usually very strong and can withstand high compressive forces

→ Any material (steel, glass, aluminium alloy, concrete, timber) is available in many different forms according to the additions, treatments, reinforcements and the dimensions and shape

| DIFFICULTY RATING | DATE |

How do changes in temperature affect construction materials

You need to know:

* how different construction materials absorb, store or transfer heat

* whether a material moves (**expands** or **contracts**) with changes of temperature.

GROUND RULES: HEAT CAPACITY (THERMAL CAPACITY)

Heat capacity is a measure of the ability of a material to absorb and store heat as temperature changes. The specific heat capacity of a material is the amount of heat energy (**Joules**) required to raise the temperature of one gram of material by one degree Celsius. For example, it takes 376J to heat a kilogram of copper by 1°C. A kilogram of polythene needs about 6 times as much heat as copper to raise the temperature by 1°C.

FIG 3.5: Examples of specific heat capacity

Substance	J/kg °C
Granite	330
Copper	376
Mild steel	502
Glass	830
Concrete	880–1040
Aluminium	920
PVC	1040
Polystyrene	1250
Standard hardboard	1250
Insulating fibreboard	1400
Perspex	1460
Timber	1500
Polyethylene	2300
Water	4187

You can see from the table that copper heats up more easily than polystyrene. Materials with a high specific heat capacity require more thermal energy to heat them up, but when they cool, the temperature falls more slowly and more heat is released.

Which would be best type of material for a storage heater?

First, find out which material absorbs the greatest amount of heat – this depends on the total mass of the material that can be packed into the heater and the specific heat capacity of the material.

The best material would have the highest value of density x specific heat capacity.

Heat energy absorbed and stored = J/m^3 = kg/m^3 x J/kg °C

A clay brick has a density of $1650kg/m^3$ and a specific heat capacity of 0.84J/kg°C.

Calculate the heat energy absorbed for this type of brick

Heat energy = Density x Specific Heat Capacity

∴ Heat energy = 1650 x 0.84 = $1386J/m^3$

NOW TRY THIS...

Which of the following materials would you select for the storage heater?

FIG 3.6: Material to select for a storage heater

Material	Density	Specific heat capacity
Concrete	2400	0.88
Glass	2580	0.84
Brick	1920	0.84

WATCH IT!

Heat loss from a building

A building loses heat when the surrounding air temperature is lower than that of a building. Heat is lost by conduction, convection, radiation and evaporation, so to be energy efficient you need to think of ways to reduce the heat loss.

[CONCEPTS]

→ **conduction** heat passes directly from the surface to the surrounding air. Conduction transfers heat through a building.

→ **radiation** infra-red radiation or heat waves are given off all surfaces. Dark surfaces radiate heat better than white surfaces.

→ **convection** air currents. Surfaces exposed to high winds will lose heat fast.

→ **evaporation** water absorbs a lot of heat during evaporation and cools surfaces. Damp surfaces lose heat quicker than dry surfaces.

Thermal movement of materials (movement caused by heat)

When the temperature changes, materials may expand or contract. Most materials expand when heated and contract when cooled. The amount of thermal movement depends on the type of material.

The coefficient of linear thermal expansion (or linear expansivity) is a measure of thermal movement for a given change in temperature. The change in length is measured in millimetres (mm). The original length is measured in metres (M). The unit for the coefficient of linear expansion is millimetres per metre °C (mm/M degrees °C).

$$\text{Coefficient of linear thermal expansion} = \frac{\text{Change in length (mm)}}{\text{original length (M)} \times \text{change in temperature (°C)}}$$

➡ **Change of state** a physical change where the chemical structure of molecules stays the same but a substance may change from solid to liquid, liquid to gas and back again, gas to liquid, liquid to solid. For example, ice to water to water vapour. Changes from solid to liquid and liquid to gas involve expansion of the molecules and this process absorbs heat (endothermic). Changes from gas to liquid and liquid to solid involve contraction of the molecules and this process releases heat (exothermic).

➡ **Enthalpy (latent heat)** the heat absorbed or released when a substance changes state, e.g.:

* heat of fusion – the heat absorbed when a substance changes from solid to liquid or released when a substance changes from liquid to solid

* heat of vaporisation – the heat absorbed when a substance changes from liquid to gas or released when a substance changes from gas to liquid.

➡ **Sensible heat** the heat you can feel. When heat is transferred by conduction you can measure a change in the temperature of the material. It is the heat absorbed by a material but which does not cause a change in state.

FIG 3.7: Coefficients of thermal expansion for construction materials in general use

Material	α	Degree of movement
Polythene	144/198	High
Acrylics	72–90	
PVC	70	
Timber – across fibres	30–70	
Lead	29	
Aluminium	24	
Polyesters	18–25	Medium 15–25
Brass	18	
Copper	17.3	
Stainless steel	17.3	
Gypsum plaster	16.6	
Concretes	10–14	
Glass	6–9	Low 1–14
Plywood	4–16	
Fired clay bricks – length	4–8	
Mortars	11–13	
Timber – longitudinal	3–6	

α = coefficient of linear thermal expansion in parts per million per Celsius degree at 20°C

WATCH IT!

Some substances have a negative expansion coefficient, such as water that expands when it freezes. The pressure of the increased volume of ice causes pipes to burst in winter.

Thermal movement can also cause:

* **cracks in large walls**

* **cracks in glass**

* **expansion and buckling of roofs**

* **fractures in pipes.**

Using and accommodating the effects of temperature change

Changes in temperature can have a major effect on certain materials, for instance by causing them to expand (get larger) or to shrink. Sometimes we can use these changes to our advantage, such as in everyday appliances like refrigerators, air conditioners, thermostats and gas ignition safety devices. But sometimes the effects of temperature change are not useful and could cause damage, such as freezing water bursting pipes. In these cases we have to find ways around the problem.

Using the expansion and contraction of metals

Thermostats

A thermostat is an electrical device used in cookers, heating systems, irons, kettles and refrigerators to control temperature. Thermostats can be set to a particular temperature; when this is reached, the thermostat automatically switches off the heater. When the temperature drops below the set temperature, the thermostat switches the heater back on.

In order to work, one type of thermostat relies on the expansion and contraction of metals in response to temperature. In this, the switch controlling the heat is operated by an electrical circuit containing a **thermostat strip** that completes the circuit, The two metals making up the strip have different **coefficients of expansion**, so one metal expands faster than the other making the strip bend, eventually breaking or completing the circuit when a particular temperature is reached.

FIG 3.8: Thermostat strip

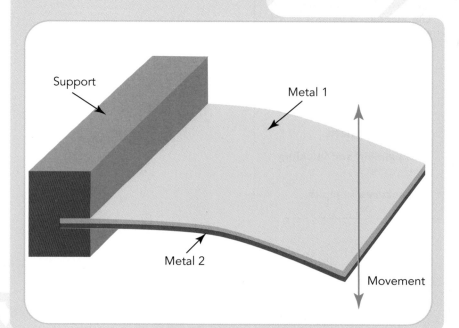

Support

Metal 1

Metal 2

Movement

IN THE REAL WORLD...

A frost thermostat protects boilers and pipework located in frost risk areas, for example, external boilers. The frost thermostat is installed next to the appliance and set to about 4°C – just above freezing. When the temperature drops this low, the frost thermostat is activated, and a feed to the motorised valve starts the pump and boiler.

Using the transfer of heat

The fact that heat transfers from one material to another can also be used to our advantage, for instance in everyday items such as radiators, boilers and refrigerators.

[CONCEPTS]

Heat flow heat flows from a hot substance to a cool substance. The hot substance gets cooler and the cool substance gets hotter until both are the same temperature. For example:

* the heat inside a house flows through the walls, windows and roof to the cooler air outside, making the inside of the house colder

* the heat inside a radiator flows into a cooler room.

Refrigeration

Without refrigeration food would go off very quickly. Refrigeration works by transferring heat from the inside of the refrigerator to the outside air. The principle behind refrigeration is the **evaporation** of liquid, when liquid evaporates it absorbs heat.

A refrigerator has a pipe circuit filled with refrigerant such as freon. The freon vapour passes into a compressor and is discharged at high pressure and heat. The freon then passes through a condenser. This has cool air or water flowing round it and absorbs heat from the vapour causing it to condense. The high-pressure liquid refrigerant then passes through a metering device, which reduces the pressure and causes the liquid to evaporate. The evaporating liquid passes directly into an evaporator, which is surrounded by the air inside the refrigerator. The heat in this air moves into the cold evaporating refrigerant. The evaporator will always be colder than the inside of the refrigerator and will keep removing heat from it, keeping it cool.

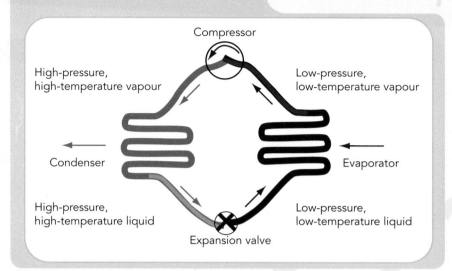

FIG 3.9: Refrigeration cycle

Accommodating the expansion and contraction of materials

Materials expand and contract due to the effects of:

* moisture

* temperature.

In the construction industry, it is important to take steps to avoid or prevent damage due to the expansion or contraction of materials.

Moisture and porosity

The construction industry uses a lot of **porous** materials. These can absorb moisture and expand. When the liquid evaporates the material will shrink or contract.

Here are four examples.

1. Dry timber expands when it absorbs water. When this moisture slowly evaporates the spaces and tubes inside the wood contract and so the timber shrinks. This is why damp or **unseasoned** timber should never be used in a building.

2. When paint films dry, the solvents evaporate from the surface and the film shrinks or contracts – dry paint film is thinner than wet film.

3. When the bottoms of wooden doors and windows are not sealed properly with paint, the timber will absorb moisture from the air and expand. This can lead to the door or window, not fitting the frame causing it to stick and be difficult to open and shut.

4. When buildings are constructed a very large amount of water is used in concrete, mortar and plaster. Over a period of time this dries out and some shrinking or contraction takes place.

Water expands on freezing

Water turns to ice at a temperature of 0°C. Ice is made of crystals. When water freezes into ice it expands and takes up more space than it did as liquid water. This expansion can cause damage to building materials. For example:

* Water pipes may freeze in winter; the expansion of the ice can force open pipe joints or split a pipe – it is important to **lag** or cover pipes to protect them from the cold.

* Bricks can absorb water and this can freeze in winter causing the brick to crack or **spall**. It is important to select the right brick in areas subject to low temperatures.

Water expands on heating

In central-heating systems the volume of water expands by about 4 per cent when heated. For instance, a system containing 100 litres will expand to 104 litres. A special cistern, called a feed and expansion cistern, must be installed in the system that has the space to take up the additional volume of heated water.

Temperature

Nearly all materials expand when heated; this is because the heat causes the molecules to move more rapidly and move further apart. Expansion and contraction of materials in response to temperature changes causes **thermal movement**.

A simple example of thermal movement is a thermometer, which uses mercury in a graded tube. As the temperature increases the mercury expands and rises up the tube; as the temperature decreases the mercury contracts and drops in the tube. But not all instances of thermal movement are this useful or even desirable.

IN THE REAL WORLD...

The following are examples of thermal movement and measures taken to avoid or prevent negative effects.

* Roofing felt expands when heated but does not contract. It is usually covered with a heat-reflecting surface layer, such as a layer of white stones or an aluminium reflecting surface.

* Plastic gutters usually show a raised fixing mark on the inside of joints, this is to allow for expansion due to heat. If gutter joints do not have expansion marks you should allow 3mm for every metre run.

* When cladding is installed, fixings are used to control movement caused by thermal expansion and contraction.

* Lead has a high coefficient of expansion at 0.0000297 for 1°C. It is important to include regular expansion joints in lead flashing to allow for thermal movement.

* Bridges, particularly metal bridges, are constructed with expansion joints built in to allow for movement.

* Long runs of pipe in hot water systems have expansion joints or bends to allow for expansion when hot water passes through.

NEED TO KNOW...

GO TO...

For information on the Web about **forces**, **loads** and **moments** see the resources on the CD accompanying this book.

HUH?

Absorbent – the ability of a material to take in and hold another substance

Coefficient of expansion – the change in length, area or volume of a material per unit change in temperature

Evaporation – the change of state of a liquid into a vapour

Lag – to insulate an area or item, such as a tank

Physical change – change of state that does not involve a chemical reaction

Porous – containing air spaces

Spall – crumbling of the weathered face of a brick or block

Thermal movement – the change in original shape or size of a material due to temperature change

Thermostatic strip – a strip made of two metals with different coefficients of expansion

Unseasoned – new timber that requires drying out

CHECKED

UNIT	The use of science and mathematics in construction	BTEC
THREE		FIRST
SUBJECT	Using and accommodating the effects of temperature change	

➔ Thermal expansion and contraction causes thermal movement

➔ Porous materials absorb water; timber expands when it absorbs water

➔ Water expands on freezing and can cause pipes to crack

➔ Water expands on heating – a central heating system needs to allow for this expansion

DIFFICULTY RATING	DATE

Solving construction problems

An equation simply says 'this equals that'

To solve the everyday calculations that come up in construction you will need to know a few basic formulae. A mathematical formula is an equation showing a standard way to solve a problem. Some of the commonest formulae used in construction include those for calculating the area and volume of different shapes.

GROUND RULES: AREA

Area refers to the size of a surface. You need to measure two **dimensions** to calculate area. The units used to describe area are always "squared", such as square centimetres or square metres. In the construction industry you may need to calculate the area of walls, windows, floors and roofs in order to work out the quantities of bricks, paint, glass or tiles you need.

HOW IT WORKS...

To calculate the area of a square or rectangle you need to multiply the length by the width.

Area of a square or rectangle = l x w.

FIG 3.10: Area of a square

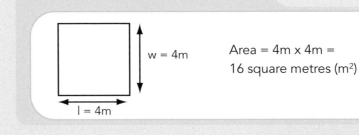

w = 4m

l = 4m

Area = 4m x 4m = 16 square metres (m²)

NOW TRY THIS...

FIG 3.11: Area of a rectangle

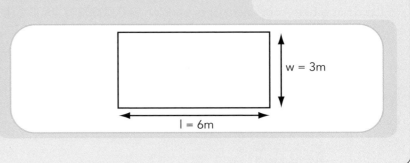

w = 3m

l = 6m

To calculate the area of a triangle

Multiply the height by half the base.

Area of a triangle = h x $\frac{b}{2}$

FIG 3.12: Area of a triangle

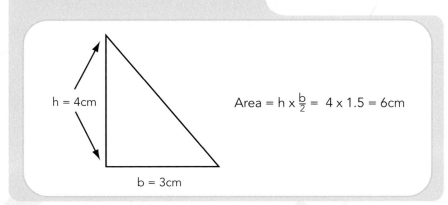

h = 4cm

Area = h x $\frac{b}{2}$ = 4 x 1.5 = 6cm

b = 3cm

To calculate the area of a trapezium

Add the lengths of the **parallel** sides, divide by two, then multiply by the height.

Area of a trapezium = $\frac{1}{2}$ (a + b) x h.

FIG 3.13: Area of a trapezium

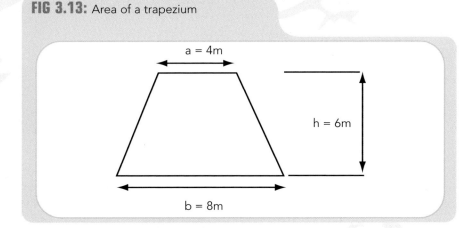

a = 4m

h = 6m

b = 8m

Area = $\frac{1}{2}$ (8 + 4) x 6 = 36 square metres

To calculate the area of a circle

Multiply the **constant**, Pi (π), by the radius of the circle, r, squared. The value of Pi is given on scientific calculators: look for the key marked π. If you do not have π on your calculator, you can get an approximate value of Pi by dividing 22 by 7.

This is written Pi $\approx$ 22 ÷ 7

When you see brackets in an equation, do that part of the calculation first. See pages 111–112 for more on brackets.

Area of a circle = πr²

FIG 3.14: Area of a circle

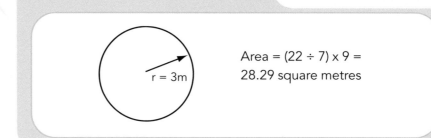

r = 3m

Area = (22 ÷ 7) x 9 =
28.29 square metres

The circumference (distance around the edge of the circle) = 2πr

GROUND RULES: CALCULATING VOLUME

Volume is the amount of space occupied by an object or substance. It is a three-dimensional quantity, expressed in cubic centimetres or metres. You need to work out volume to:

* measure the amount of water, cement and sand required to make enough concrete to fill a given space

* find the volume of a room to be heated

* calculate how much earth must be moved when forming foundations and trenches.

HOW IT WORKS...

To calculate the volume of a cube or **cuboid**, multiply the length by the width by the height.

Volume = l x w x h

FIG 3.15: Volume of a cube

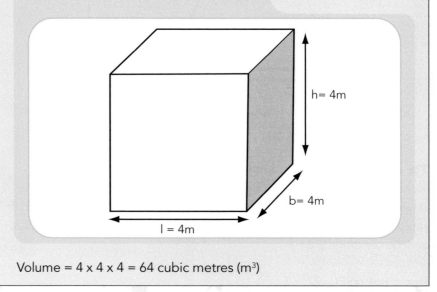

h= 4m

b= 4m

l = 4m

Volume = 4 x 4 x 4 = 64 cubic metres (m³)

NOW TRY THIS...

Calculate the volume of a cuboid of length 8cm, width 4cm and height 3cm.

To calculate the volume of a cylinder

Multiply the area of the circle at one end by the height of the cylinder.

Volume = πr² x h

FIG 3.16: Volume of a cylinder

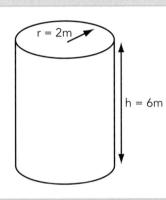

Volume = (22 ÷ 7) x 2² x 6
= (22 ÷ 7) x 4 x 6 =
75.43 cubic metres (m³)

GROUND RULES: PERCENTAGES

A percentage is effectively a fraction expressed as parts of a hundred. In other words, a percentage is a fraction where the **denominator** is always 100. A half, for instance, would be expressed as 50 per cent. You need to know how to work out percentages so that you can make adjustments to material quantities and costs. We use the sign % (per cent) to show that, for example, 46% means '46 parts out of one hundred'.

HOW IT WORKS...

What is 40% of £18,000?

Divide £18,000 into 100 parts. Each part (or 1%) is £180. Now multiply this by 40 = £7,200

Here is another example.

I have a loan of £200 to buy a new concrete mixer, and I have to pay £230 back – what is the interest rate?

The additional payment is £230 – 200 = £30

1% of £200 is £2

Therefore the interest rate percentage is 30 divided by 2 = 15%

NOW TRY THIS...

I have a builder's bill of £58 for timber and nails, the nails cost 18% of the total. How much is the cost of the nails?

Transposing formulae

You will use mathematical formulae over and over again in your construction career, but in order to find out what you need to know, you may have to turn a formula around to fit the particular real-life problem you are dealing with. This is called **transposing**.

GROUND RULES: TRANSPOSING

Suppose you need to work out how much of a full can of paint you have used. You know that the amount of paint in a full can (X) will be equal to the amount of paint used (Y) plus the amount left in the can (Z)

$$X = Y + Z$$

If a can of paint holds 5 litres (X) and you have 3 litres left (Z), you know you have used 2 litres (Y). You have rearranged the formula in your head to work out what Y is.

The amount used (Y) = the amount in a full can (X) – the amount left (Z)

$$Y = X - Z$$

HOW IT WORKS...

There is an important mathematical rule to remember: **if you change one side of the equation, you must change the other side in exactly the same way.**

When you worked out what Y was, you subtracted Z from both sides of the original equation:

$$X = Y + Z$$

$X - Z = Y + Z - Z$ on this side of the equation, the plus Z and minus Z cancel each other out.

X – Z = Y This is exactly the same as Y = X – Z

When solving equations you can:

* **add** the same quantity to both sides

* **subtract** the same quantity from both sides

* **multiply** both sides by the same quantity

* **divide** both sides by any non-zero quantity.

Think of an equation as a pair of scales where each side of the scales (the equals sign, =) must be balanced.

For instance, to transpose the formula Z = X + Y + 2 to make Y the subject (the value you want to find):

$$Z = X + Y + 2$$

$$Z - X - 2 = X + Y + 2 - X - 2$$

$$Y = Z - X - 2$$

Here is another example, transpose X = Y x 2 to make Y the subject.

You have to divide both sides by Z to leave Y alone on one side of the equation.

$$\frac{X}{Z} = \frac{Y \times Z}{Z} \qquad\qquad \text{Z divided by Z = 1}$$

$$Y = \frac{X}{Z}$$

NOW TRY THIS...

V = T x R   transpose to make R the subject

S = V x W   transpose to make W the subject

P = Q² x R   transpose to make Q the subject.

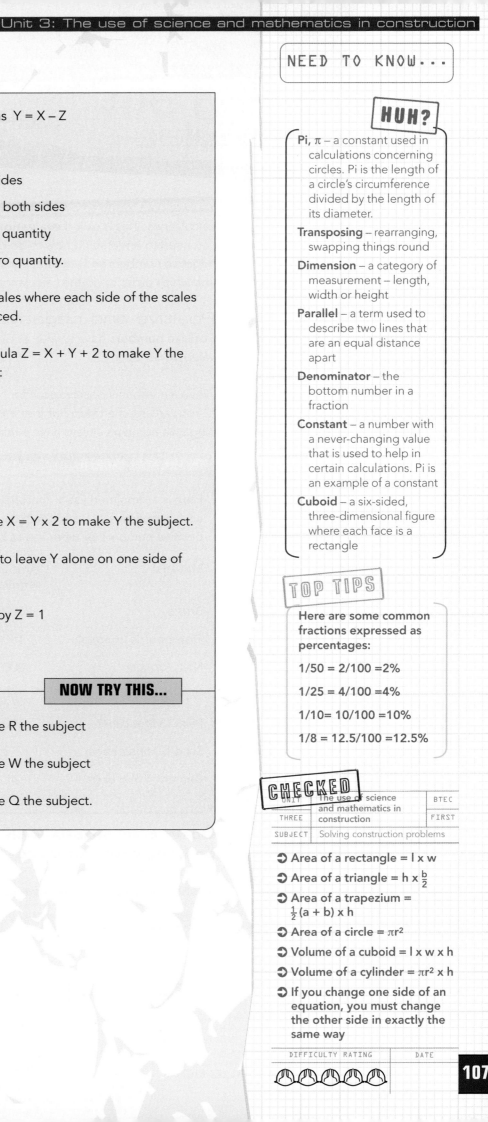

NEED TO KNOW...

HUH?

Pi, π – a constant used in calculations concerning circles. Pi is the length of a circle's circumference divided by the length of its diameter.

Transposing – rearranging, swapping things round

Dimension – a category of measurement – length, width or height

Parallel – a term used to describe two lines that are an equal distance apart

Denominator – the bottom number in a fraction

Constant – a number with a never-changing value that is used to help in certain calculations. Pi is an example of a constant

Cuboid – a six-sided, three-dimensional figure where each face is a rectangle

TOP TIPS

Here are some common fractions expressed as percentages:

1/50 = 2/100 =2%

1/25 = 4/100 =4%

1/10= 10/100 =10%

1/8 = 12.5/100 =12.5%

CHECKED

UNIT	The use of science and mathematics in	BTEC
THREE	construction	FIRST
SUBJECT	Solving construction problems	

➲ Area of a rectangle = l x w

➲ Area of a triangle = h x $\frac{b}{2}$

➲ Area of a trapezium = $\frac{1}{2}$ (a + b) x h

➲ Area of a circle = πr²

➲ Volume of a cuboid = l x w x h

➲ Volume of a cylinder = πr² x h

➲ If you change one side of an equation, you must change the other side in exactly the same way

DIFFICULTY RATING	DATE

Working out formulae

You work out formulae by putting in the values you know and doing the calculations. This is called **evaluation** of formulae. To evaluate formulae you need to understand how to do calculations with positive and negative numbers and with fractions, how to work out values in brackets and which order to do the calculations in.

Positive and negative numbers

Positive numbers have a value above zero. **Negative numbers** have a value below zero.

When a number is positive you don't need to put a plus (+) sign in front of it. You can assume that a number is positive if it has no sign in front. Negative numbers always have a minus (–) sign in front:

IN THE REAL WORLD...

Here is a simplified list of materials required for a job, the amount of each supply in stock and the balance, which shows what's left (a positive number), or what has to be ordered (a negative number).

Material	Amount required	Supplies in stock	Balance
Bricks	2000	1500	–500
Floorboards (metres)	150	200	50
Nails (boxes)	1000	900	–100
Varnish (2.5 litre cans)	100	200	100
Paint (5 litre cans)	200	175	–25
Sand (1 tonne bags)	12	35	23
Cement (25kg bags)	42	196	154

Doing calculations with positive and negative numbers

The easiest way to visualise this is to use a number line with negative numbers below zero and positive numbers above zero.

FIG 3.17: Number line

For addition you move from left to right: $1 + 3 = 4$ $-3 + 5 = 2$

For subtraction, you need to move from right to left: $3 - 5 = -2$
$-1 - 3 = -4$

Rule: in multiplication and division calculations

+ and + = + $+5 \times +7 = 35$ $+25 \div (+5) = 5$

− and − = + $-5 \times -7 = 35$ $-25 \div (-5) = 5$

+ and − = − $+5 \times -7 = -35$ $+25 \div (-5) = -5$

Fractions

Fractions are a way of expressing how many parts of a whole number we have left if we divide it by another number. If you divide one brick into four equal parts and take three parts away, you're left with one part out of the original four; or in other words, a quarter.

Simple fractions are numbers between nought and one and cannot have more units on the top than on the bottom. Divide the top number by the bottom number to get the correct answer as a simple fraction.

Examples

$\frac{5}{4} = 1\frac{1}{4}$ $\frac{9}{4} = 2\frac{1}{4}$ $\frac{17}{5} = 3\frac{2}{5}$

Cancelling

When working out fractions, cancel down numbers as much as possible. You do this by dividing the top and bottom numbers by the same amount.

Examples

$\frac{10}{100}$ divide the numerator and denominator by 10: $\frac{1}{10}$

$\frac{8}{16}$ divide the numerator and denominator by 8: $\frac{1}{2}$

$\frac{6}{15}$ divide the numerator and denominator by 3: $\frac{2}{5}$

GROUND RULES: DOING CALCULATIONS WITH FRACTIONS

$$1 \div 4 = \frac{1}{4}$$

The top number is called the numerator and tells you how many parts there are.

The bottom number is called the denominator and tells you how many equal parts there were in the whole object.

HOW IT WORKS...

To add or subtract fractions with equal denominators, you just add or subtract the numerator.

$$\frac{1}{2} + \frac{1}{2} = \frac{2}{2} = 1 \qquad \frac{5}{8} - \frac{2}{8} = \frac{3}{8}$$

If the denominators are not the same, you have to find a **common denominator** that is a number that both denominators will divide into leaving no remainder.

For example, you can't add a half to a quarter, but you can call a half two quarters and then add the two quarters to the one quarter, making three quarters.

$$\frac{1}{2} + \frac{1}{4} = \frac{2}{4} + \frac{1}{4} = \frac{3}{4}$$

The common denominator is 4.

Note that 2 goes into 4 twice, therefore you have to multiply the nominator by two as well:

$\frac{1}{2}$ becomes $\frac{2}{4}$

Subtracting fractions works the same way:

$$\frac{1}{2} - \frac{1}{4} = \frac{2}{4} - \frac{1}{4} = \frac{1}{4}$$

When multiplying fractions, you can think of the multiplication sign as meaning 'of'.

For instance:

7 x 3   means seven sets of three

$\frac{1}{2}$ x $\frac{1}{4}$ means a half of a quarter

To multiply fractions, you just multiply the numerators and then multiply the denominators.

Example

$\frac{1}{2} \times \frac{1}{4} = \frac{1}{8}$ $(1 \times 1 = 1)$ A half of a quarter is an eighth.
$(2 \times 4 = 8)$

When you divide fractions, you are working out how many times one fraction 'goes into' another. To do this you turn the second fraction upside down and then multiply.

$\frac{1}{2} \div \frac{1}{4}$ becomes $\frac{1}{2} \times \frac{4}{1}$

The answer is then: $\frac{1}{2} \times \frac{4}{1} = \frac{4}{2} = 2$

NOW TRY THIS...

1. Find out the volume of concrete required for three small jobs.

   Imagine you are a surveyor preparing a quote. You have to work out the correct amount of concrete to pour bases 500mm deep on three small plots of land:

   * plot A = $\frac{3}{4}$m x $\frac{1}{2}$m
   * plot B = $\frac{1}{4}$m x $\frac{3}{4}$m
   * plot C = $\frac{1}{2}$m x $\frac{1}{2}$m.

   You need to work out the size of each area and then add the areas together to produce the sum total of the area of the three plots.

   What is the total volume of concrete required for the three plots?

2. Calculate the total amount of timber wastage at the end of a job.

   At the end of a job, the following timber is left over:
   * 30 lengths of 3/5m of softwood
   * 20 lengths of hardwood measuring $\frac{1}{4}$m in length each.

   What is the total amount of timber wastage at the end of the job?

GROUND RULES: BRACKETS

Brackets are used to keep part of a formula together, especially where the contents of the bracket have to be worked out and then multiplied or divided by another value.

HOW IT WORKS...

The area of a trapezium is $\frac{1}{2}$ (a + b) x h.

Work out the bracket first; suppose a = 4, b = 6:

$\frac{1}{2}(4 + 6) = \frac{1}{2}(10) = 5$

This could also be written as $(\frac{1}{2} \times 4) + (\frac{1}{2} \times 6) = 2 + 3 = 5$

If there are two brackets next each other, you need to multiply them together.

$(5+5)(4+4)$ means $(5+5) \times (4+4) = 10 \times 8 = 80$

If an equation has brackets inside other brackets, work out the inside brackets first.

$3(2 + (5 \times 2))$ means $3 \times (2 + 10) = 3 \times 12 = 36$

Rule: If equations have brackets, work these out first, starting with the inside brackets and including any multiplication or division of the contents of the brackets.

NOW TRY THIS...

1. How many patio slabs do you need for a job?

   A client wants two patio areas, one in the front and one in the back garden. The garden measurements are:
   – back garden: 6 slabs by 4 slabs
   – front garden: 8 slabs by 3 slabs.

   You decide to add 2 slabs for each area to allow for wastage.

   Create an equation using brackets to show how to work out the total number of slabs – then work out how many slabs you need.

2. Calculate the total number of 'worker days' for four jobs.

   Note: A 'worker day' is one worker x one day.

   You are a manager of a construction company and you need to work out how many worker days your company needs to commit to four new jobs. The requirements assessed for each job are:
   * job A = 3 workers for 8 days
   * job B = 5 workers for 7 days
   * job C = 2 workers for 14 days
   * job D = 6 workers for 3 days.

   You decide to add one worker day for each job to allow for **contingencies**.

   Create an equation using brackets to show how to work out the total number of worker days – then work out how many worker days you need.

3. How much will it cost to hire a cement mixer?

Hiring a cement mixer costs £60 deposit then a further £35 per day.

How much does it cost to hire the cement mixer for 9 days?

4. What is the total cost?

Adam buys three silicon applicator guns for £9.99 each and 13 tubes of sealant at £3.98 each.

How much does he pay altogether?

WATCH IT!

The order of events

You have to do your calculations in the right order otherwise you get the wrong answer.

One way to remember is to use the acronym BODMAS, this stands for:

Brackets

Order (numbers squared or cubed)

Division

Multiplication

Addition

Subtraction.

NEED TO KNOW...

HUH?

Common denominator – in fractions a number that divides into all other denominators

Contingencies – things that stop work going to plan – such as worker illness

Evaluation – putting a value on something. When you put the numerical values into formulae, you can find the numerical solution

Negative number – less than zero

Positive number – greater than zero

CHECKED

UNIT	The use of science and mathematics in construction	BTEC
THREE		FIRST
SUBJECT	Working out formulae	

➲ To add or subtract fractions with equal denominators, you just add or subtract the numerator

➲ To multiply fractions, multiply the numerators and then multiply the denominators

➲ To divide fractions, turn the second fraction upside down and then multiply

➲ If equations have brackets, work these out first – start with the inside brackets and include any multiplication or division of the contents of the brackets

DIFFICULTY RATING	DATE

Solving some real construction problems

Use the information from the previous pages to solve these real construction problems.

Mensuration techniques

Mensuration means finding out values, such as length, height, width, area or volume, by measurement and calculation. For example, a wall is 4 metres long and 2.5 metres high; it has a window 1 metre by 1.5 metres. What is the area of the wall surface? Obviously, the window is not part of the wall surface so you need to subtract the area of the window from the total wall area.

Answer: $(4 \times 2.5) - (1 \times 1.5) = 8.5m^2$

WATCH IT!

✳ Check your measurements for accuracy – measure twice.

✳ Check the units you are working with – convert all measurements to the same unit. For instance, to find the area of a window 75cm x 1.2m, convert to the same unit first:
0.75 x 1.2m
or
75 x 120cm.

TRY THIS

Calculate the cost of painting a wall

To calculate the cost of painting the interior wall mentioned above, you need to know:

✱ the walls have to be painted twice

✱ a tin of paint costs £23.99

✱ each tin covers $62m^2$.

How many tins of paint will be needed?

Wall surface area x 2 coats = _____m^2 x 2 coats =_____m^2 total coverage

Total coverage _____ $m^2 \div 62m^2$ = _____ tins of paint required

Calculate the cost.

_____tins x £23.99 per tin = _____

Calculating how much concrete you need for foundations

Before you can build walls, a base has to be laid to form the foundations of the building. The foundations ensure the building will not move. There are two types of foundation, one around the perimeter, which is deeper to take the weight of the structure, and an infill or raft foundation. The base is like a raft and is formed by digging out a trench and filling it in with concrete reinforced with steel mesh or bars.

FIG 3.18: Cross-section of the foundations for the base

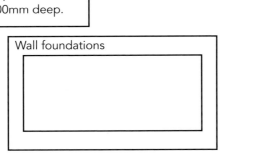

This is the edge of the base, this foundation would be 26m in length, 1 metre in width, and 900mm deep.

Raft is 300mm deep

Wall foundations

TRY THIS

If the trench for the foundation of a wall is 26m long and 1m wide, and the thickness of the concrete is 200mm. What volume of concrete is required?

Volume of concrete   =   length x width x thickness

=   _____ x _____ x _____

=   _____m³

Calculate adjustments to quantities

You may need to make adjustments to prices or the amount of materials you need because of:

* Value added tax (VAT) – many builders' merchants display prices excluding the cost of VAT, so you need to add this to prices displayed to find the overall cost. VAT currently stands at 17.5 per cent.

* Breakage and wastage of materials on the job – when ordering materials, you need to make an allowance for cutting or breakage, this allowance is usually 40 per cent.

* Allowance for dampness of materials, such as sand.

* Allowance for shrinkage of concrete.

Calculate the cost of applying interior wall finish

You can produce a good quality finish on the interior of a room if you dot and dab plasterboards to the walls.

To calculate the cost of applying plasterboard to a room 26m x 10m you need to find:

* the number of plasterboards required

* the cost of the plasterboards plus VAT

* the cost of bonding adhesive plus VAT.

Note that in this case we are not worrying about the height as it works out at 2.4m, the height of a plasterboard. In other instances you may have to take height into consideration.

How many plasterboards do you need?

You need to calculate the perimeter of the room and divide the answer by the width of the plasterboards. The plasterboards are 1200mm wide and 2400mm high.

Remember to convert the unit measurements of the boards to the same units as the wall measurements: 1.2m wide x 2.4m high.

Perimeter   =   2 x length of room + 2 x width of room

              =   _____ m + _____ m =   _____ m

Number of boards = perimeter ÷ width of boards

              = _____m ÷ 1.2m = _____ boards

How much will the boards cost?

One plasterboard costs £3.99 + VAT. Assume that VAT is 17.5 per cent.

Cost of boards = number of boards x (£3.99 + 17.5%). Remember to work out the brackets first.

= _____ x _____ = £ _____

How much will the bonding adhesive cost?

The bonding adhesive used to stick the boards to the walls is £4.99 + VAT per bag. Each bag will attach three boards.

The number of bags required = number of boards ÷ 3 = _____

Cost of bonding adhesive =
number of bags required x (£4.99 + VAT) = £ _____

Total cost of applying interior wall finish = cost of boards + cost of bonding adhesive

= £ _____

Calculate the cost of materials to be ordered

Ordering up the right amount of materials is an essential skill. If you order too much you have wasted money, and if you order too little you won't be able to complete the job. Timber is one of the commonest materials you will need to calculate. It is used structurally and for items such as floorboards, window and door frames, skirting boards and staircases.

TRY THIS

A construction job is estimated to need 65 metres of 100mm x 50mm timber (four by two, 4" x 2"). You need to work out the allowance (10 per cent) and the total cost of the timber. The timber is sold in lengths and the cost per linear metre is £2.65 excluding VAT.

You need to work out:

* the 10% allowance = _____ metres

* the total amount of timber = _____ metres

* the cost of the timber excluding VAT = £_____

* the cost of the timber including VAT (17.5%) = £_____

Bulking calculations

When buying sand, you need to think about whether it is wet or dry – this will depend on weather conditions. In damp weather, surface water on sand particles increases the volume and weight of the sand. Sand is no longer sold by weight but in one-size bulk bags. You need to make an allowance for the bulking effect of moisture when you work out how many bags of sand are required for a job. To be absolutely sure that you have enough sand allow 33 per cent for bulking.

Calculate the shrinkage of concrete

Concrete shrinkage is caused by loss of moisture from the paste. It is influenced by a variety of factors, including:

* environmental conditions such as temperature and humidity

TRY THIS

A job requires 30 bags of dry sand. The cost of a bulk bag is £31 plus VAT.

In order to ensure that you have enough materials you must increase your order by 33 per cent.

Calculate the total cost of the sand you need to buy, including VAT.

* the size of the surface area you are working – sometimes called the member

* factors associated with concrete material, such as volume of aggregate and water to cement ratio.

TRY THIS

Calculate the volume of a concrete base after shrinkage. Make an allowance of 25 per cent for concrete shrinkage.

You have laid a base 500mm deep, 25m long and 12m wide:

* calculate the volume of concrete laid

* calculate the volume after shrinkage.

Calculate proportions

Materials such as concrete and mortar are made by mixing ingredients by proportion.

Concrete

Concrete is a mixture of cement, aggregate, sand and water in controlled proportions by volume. Concrete mixes are often quoted by a ratio such as:

* 1:2:4

* $1m^3$ of cement to $2m^3$ of sand to $4m^3$ of coarse aggregate.

For instance, if you put in three shovels full of cement, how much sand and aggregate will be required for the mixture?

Ratio = 1:2:4

You have three shovels of cement, so multiply the ratio by three.

= 3 cement : 6 sand : 12 aggregate.

TRY THIS

If you need $10m^3$ of concrete, in the ratio 1:2:4, how much of each material will be required?

If you use 8 shovels of sand to make a concrete mix, in the ratio 1:2:4, how much cement and aggregate will you need?

Bedding mortar mix

A bedding mortar mix is made from cement and sharp sand in the ratio of 1:5. For instance, if you use 10 shovels of sand how much cement will you need for the bedding mix?

___ cement : 10 sand

Ten shovels of sand is twice the amount given in the ratio (5).

So you need twice the amount of cement as well, 1 x 2 = 2.

You need 2 shovels of cement to 10 of sand.

TRY THIS

If you use 15 buckets of sand to make a mortar mix. How much cement should you use?

Working with trigonometry – the mathematics of triangles

In construction problem solving, you will often use the special relationships between the angles and sides of right-angled triangles to calculate heights, lengths and angles, for example when dealing with roof slopes.

Angles

Angle the size of the opening between two lines measured in degrees.

A 90° angle (Figure 3.19) is called a *right angle* and the angle is always represented on diagrams as square. Other angles are shown as sectors of a circle. Angles less than 90° are called *acute* angles. A 45° (Figure 3.20).

FACT IS...

There are 360° in a circle.

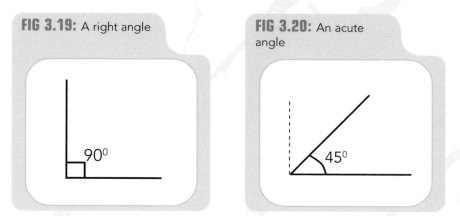

FIG 3.19: A right angle

90⁰

FIG 3.20: An acute angle

45⁰

Angles between 90° and 180° are called *obtuse* angles. The **protractor** in Figure 3.21 is measuring an obtuse angle of 140°.

FIG 3.21: An obtuse angle

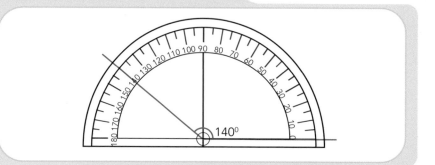

140⁰

There are 360° in a circle and 180° in half a circle. A straight line is like the diameter of an imaginary circle, so on one side of any straight line any angles should add up to 180°.

If you look at the protractor in Figure 3.22, you can see that angle (a) is 40° and angle (b) is 140°. The two angles add up to 180° in total as 40° + 140° = 180°.

FIG 3.22: Adding angles

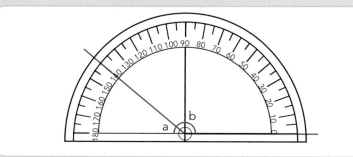

The angle sum of a triangle

Triangles have three sides and three angles. The three angles always add up to 180°. Therefore if you are given the value of two of the angles, you can easily find the third:

* if one angle is 40° and another is 60°, then the third must be:
  180 – (40 + 60) = 80°

* if one angle is 83° and another is 45°, then the third must be:
  180 – (83 + 45) = 52°.

There are five types of triangle:

* **scalene** – every angle is less than (<) 90°

* **right-angled** – one angle is exactly 90°

* **obtuse** – one angle is greater than (>) 90°

* **isosceles** – two angles and two sides are equal

* **equilateral** – all the sides are the same length and all the angles are exactly 60°.

FIG 3.23: Five types of triangle

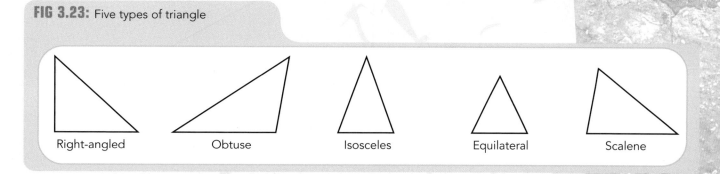

Right-angled   Obtuse   Isosceles   Equilateral   Scalene

Pythagoras' theorem

In a right-angled triangle, the square on the hypotenuse is equal to the sum of the squares on the other two sides.

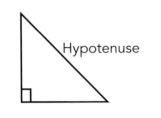

FIG 3.24: Right-angled triangle showing position of hypotenuse

Put simply, if you know the length of two of the sides, you can find out the length of the third.

Remember that a right angle is often represented as a square in the corner of the triangle? Well, the inside corner points to the longest side, the **hypotenuse**. The hypotenuse is usually called (A) and the other two sides (B) and (C).

The formula representing Pythagoras' theorem is:

$A^2 = B^2 + C^2$.

You can use Pythagoras' theorem to check that the setting out for the sides of a building are square and straight. If one side wall is 30 metres long and the other is 40 metres long, then you can work out the length of the hypotenuse and measure it:

$A^2 = 30^2 + 40^2 = 900 + 1600 = 2500$

$A = 50m$

You will find more on this in *Putting it all together to solve some real construction problems* on page 130.

Finding angles using sines, cosines and tangents

In right-angled triangles, the relationships between angles and sides can be used to find the size of an angle using special tables called sine, cosine and tangent tables.

FIG 3.25: Right-angled triangle showing the positions of the hypotenuse, opposite and adjacent sides in relation to the angle you want to find, θ

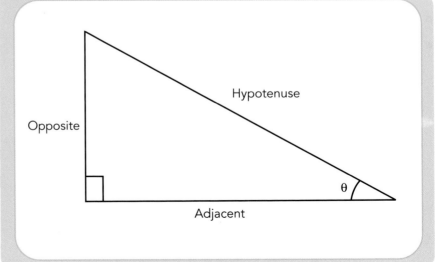

Here are the formulae for working out the sine, cosine or tangent of an angle:

To find this:	Work out this:	Memory aid
Sine θ	$\dfrac{\text{opposite}}{\text{hypotenuse}}$	SOH
Cosine θ	$\dfrac{\text{adjacent}}{\text{hypotenuse}}$	CAH
Tangent θ	$\dfrac{\text{opposite}}{\text{adjacent}}$	TOA

We use a calculator with special functions to work out the formula.

Examples: Find the value of the angle, θ, as shown in the diagrams in Figs 3.26, 3.27 and 3.28.

1. We know the length of opposite and adjacent sides, so we use tangent formula (TOA). The shaded box below shows the calculator function for use with this formula.

FIG 3.26: Tangent calculation

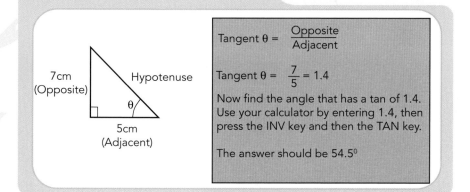

$$\text{Tangent } \theta = \frac{\text{Opposite}}{\text{Adjacent}}$$

$$\text{Tangent } \theta = \frac{7}{5} = 1.4$$

Now find the angle that has a tan of 1.4. Use your calculator by entering 1.4, then press the INV key and then the TAN key.

The answer should be 54.5^0

2. We know the length of the opposite side and the hypotenuse, so we use the sine formula (SOH).

FIG 3.27: Sine calculation

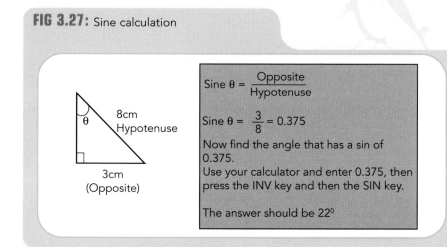

$$\text{Sine } \theta = \frac{\text{Opposite}}{\text{Hypotenuse}}$$

$$\text{Sine } \theta = \frac{3}{8} = 0.375$$

Now find the angle that has a sin of 0.375. Use your calculator and enter 0.375, then press the INV key and then the SIN key.

The answer should be 22^0

3. We know the length of the adjacent side and the hypotenuse, so we use the cosine formula (CAH).

FIG 3.28: Cosine calculation

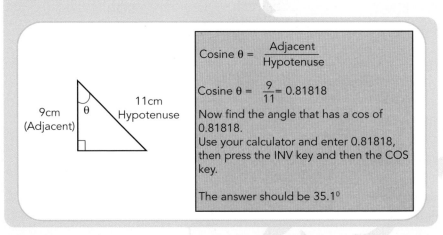

FIG 3.28: Cosine calculation

$$\text{Cosine } \theta = \frac{\text{Adjacent}}{\text{Hypotenuse}}$$

$$\text{Cosine } \theta = \frac{9}{11} = 0.81818$$

Now find the angle that has a cos of 0.81818.
Use your calculator and enter 0.81818, then press the INV key and then the COS key.

The answer should be 35.1°

Using sine, cosine and tangent

Here's an example of how you might use these calculations in a real-life situation, designing a staircase. When designing and making a staircase it is a legal requirement that the angle of the stairs is not too steep.

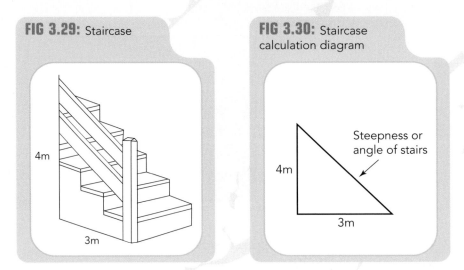

FIG 3.29: Staircase

FIG 3.30: Staircase calculation diagram

The angle of the stairs depends on the length and height of the staircase – in this case:

* the height of the stairs is 4m

* the length is 3m.

Using the sine, cosine and tangent rule, we can calculate the angle of the stairs and the remaining length.

$$\text{Tan} = \frac{\text{opposite}}{\text{adjacent}} = \frac{4}{3} = 1.33$$

Tan θ⁻¹ 1.33

The angle of the stairs is = 53.1°

TRY THIS

Calculate the angle of these three staircases:

Staircase 1: height 3.7m, distance from wall 3.2m.

Staircase 2: height 4.2m, distance from wall 3.72m.

Staircase 3: height 3.25m, hypotenuse 6m.

Finding the length of sides using sines, cosines and tangents

If you know one of the angles, and the length of one side, you can transpose the relevant formula to find the length the other sides.

Example: Find the length of the side, X.

We know the size of the angle, and the length of the adjacent side. We want to find the length of the opposite side, so we select the tangent formula (TOA)

$$\text{Tangent } 25° = \frac{\text{opposite}}{\text{adjacent}} = \frac{X}{5}$$

You can transpose this to make X the subject:

$$X = \text{tangent } 25° \times 5\text{cm}$$

Using your calculator, enter 25 and then press the TAN key. This will give you 0.4663. Now multiply this by 5. The answer should be 2.332.

The length of the opposite side, X, is 2.33 cm.

TRY THIS

On this sectional view of half of a roof truss, the 'run' is the adjacent side. What is the length of the run?

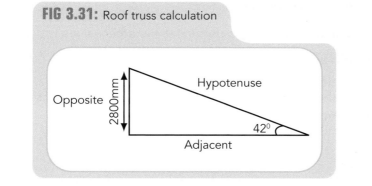

FIG 3.31: Roof truss calculation

NEED TO KNOW...

GO TO...

For information on the Web about **the use of trigonometry** see the resources on the CD accompanying this book.

HUH?

Acute – an angle less than 90°

Equilateral – a triangle in which all the sides are the same length and all angles are exactly 60°

Hypotenuse – the longest side of a right-angled triangle, opposite the right angle

Isosceles – a triangle with two angles and two sides the same size

Obtuse – an angle greater than (>) 90° and less than (<) 180°

Protractor – a transparent plastic half circle used for measuring angles

Scalene – a triangle in which every angle is less than 90°

Right-angled triangle – a triangle where one angle is exactly 90°

CHECKED

UNIT	The use of science and mathematics in	BTEC
THREE	construction	FIRST
SUBJECT	Working with trigonometry – the mathematics of triangles	

➔ There are 360° in a circle and 180° in half a circle

➔ Pythagoras' theorem: the square on the hypotenuse (A) is equal to the sum of the squares on the other two sides: $A^2 = B^2 + C^2$

➔ $\text{Sine } \theta = \dfrac{\text{opposite}}{\text{hypotenuse}}$

➔ $\text{Cosine } \theta = \dfrac{\text{adjacent}}{\text{hypotenuse}}$

➔ $\text{Tangent } \theta = \dfrac{\text{opposite}}{\text{adjacent}}$

DIFFICULTY RATING	DATE

Using graphs to work out construction problems

You can use graphs to work out construction calculations and estimates, such as:

* labour and material costs

* quantities of materials required per area or volume.

* time required for tasks

* power usage.

GROUND RULES: PLOTTING AND READING GRAPHS

A graph shows the relationship of one value to another. One value is marked on a vertical line (the y axis) and the other on a horizontal line (the x axis).

HOW IT WORKS...

You can create a graph that shows how many litres of paint you need to cover an area of wall. The table below shows how much paint you need to apply two coats to a smooth wall surface.

Square metres of wall	Litres of paint required
5	0.9
10	1.8
15	2.7
20	3.6

To plot the graph:

1. Find the first value on the y axis (5m^2).

2. Follow the horizontal line across untill you come to a point above the correct value on the x axis, 0.9.

3. Put your first dot here.

4. Continue to place the dots for all the other values.

5. Join the dots.

The values on the x and y axes for each dot are known as **Cartesian coordinates** (x, y).

FIG 3.32: A simple graph

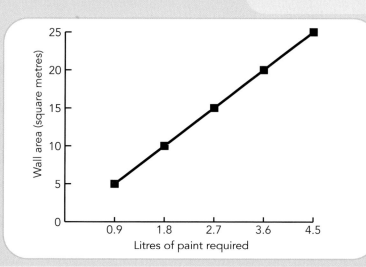

Reading a graph

On this graph, you can look up the number of square metres that you have to paint and follow the line across untill it meets the plotted line. If you follow that point down to the x axis, you will find the number of litres of paint you need to buy.

Interpolation means looking up values between the dots. The line connecting the dots shows you the relationship between x and y for every other value, up to the value of the last dot. If you want to find out how many litres of paint you need for 11.25 square metres of wall, just look up 11.25 on the y axis, follow the horizontal line across (use a ruler to make sure you stay on the same line) until you reach the plotted line, then follow the vertical line down to give you the value on the x axis.

NOW TRY THIS...

How many rolls of wallpaper do you need?

You have to wallpaper a room 2.5m high, 4m wide and 5m long.

You can get four 2.5m drops from one standard roll of normal wallpaper. The table below shows how many rolls of wallpaper you need for rooms with different perimeters.

Perimeter (m)	7	8.5	10	11.5	13	14.5	16	17.5	19
Rolls required	4	5	5	6	7	8	8	9	9

Plot the graph using the information from the table.

Calculate the perimeter of the room.

How many rolls of newspaper do you need to paper the whole room?

WATCH IT!

Remember when calculating wall area to ignore doors and windows.

127

An equation shows the relationship between different values. A graph also shows the relationship between x and y. So a graph must correspond to an equation.

HOW IT WORKS...

When x has the same numerical value as y (x = y) you get a straight-line graph like the left-hand graph in Figure 3.33. When y is equal to a multiple of x (y = mx) you get a straight line graph like the right-hand example shown in Figure 3.33.

FIG 3.33: Straight line graphs

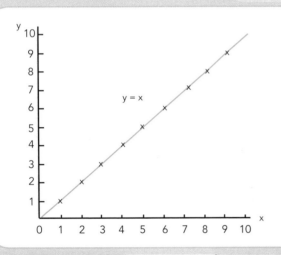

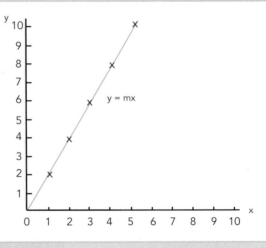

NOW TRY THIS...

Plot the two lines for resistor A and resistor B.

Data for resistor A		Data for resistor B	
Current (A)	Voltage (V)	Current (A)	Voltage (V)
0	0	0	0
5	5	2	4
10	10	3	8
15	15	4	12
20	20	5	16

What is the equation of the graph for resistor A?

What is the equation of the graph for resistor B?

TRY THIS

A quantity surveyor would use a graph to give estimated prices for specific jobs, such as the **rendering** of surfaces to exterior walls. The surveyor would calculate the work area and apply the unit price per m².

The unit price for rendering exterior surfaces is £35 per m².

Plot the graph where:

y = the unit cost per m²

x = area of wall in m²

x	1	2	3	4	5	6	7	8	9	10	11	12
y	35	70	105	140	175	210	245	280	315	350	385	420

What is the equation of the graph?

Extrapolation means finding values beyond the scope of the graph. To find out how much it would cost to render 22m² of wall, you can just obtain the values for 10m² and 12m² and add them together.

How much would it cost to render 43m² of wall?

The gradient

The gradient is the steepness of the graph. It is equal to the factor multiplying x.

* For the graph of y = x, the gradient is 1 (1y = 1 x).

* For the graph of y = mx, the gradient is m.

* For the graph of y = mx + c, the gradient is also m; the steepness of the line is the same as y = mx even though the lines cut the y axis at different points.

The y intercept

The plotted line meets the y axis when x = 0. This is called the y intercept.

* For the graphs of y = x and y = mx, the intercept is at y = 0. When x = 0, y = 0.

* For the graph of y = mx + c, the y intercept will be at c; when x = 0, y = c.

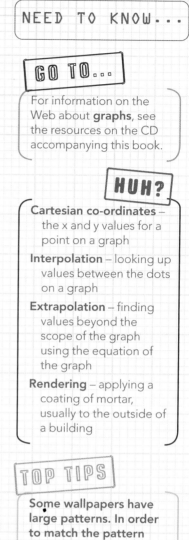

NEED TO KNOW...

GO TO...

For information on the Web about **graphs**, see the resources on the CD accompanying this book.

HUH?

Cartesian co-ordinates – the x and y values for a point on a graph

Interpolation – looking up values between the dots on a graph

Extrapolation – finding values beyond the scope of the graph using the equation of the graph

Rendering – applying a coating of mortar, usually to the outside of a building

TOP TIPS

Some wallpapers have large patterns. In order to match the pattern you will have to waste some paper. Consult the manufacturer or supplier about how many drops per roll you will get.

CHECKED

UNIT	The use of science and mathematics in construction	BTEC
THREE		FIRST
SUBJECT	Using graphs to work out construction problems	

➜ A graph shows the relationship of one value to another

➜ The gradient is the steepness of the graph and is equal to the factor multiplying x

➜ The y intercept – where the plotted line meets the y axis (x = 0)

DIFFICULTY RATING	DATE

You can use your knowledge of trigonometry and graphs to solve construction problems such as:

* planning and designing a staircase or **pitched roof**

* calculating length, width and angles

* setting out buildings

* making sure planes (walls, floors) are **aligned** and square

* using data to find the relationship between **variables**

* extrapolating and interpolating values for variables.

Plan and design a staircase

FIG 3.34: A straight staircase

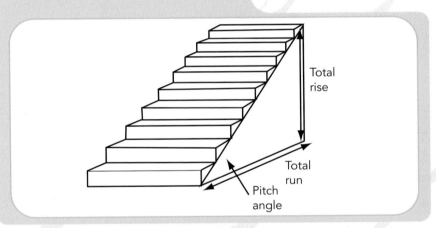

A straight staircase has the overall shape of a right-angled triangle. Each step is also two sides of a right-angled triangle:

* the flat part of each step is called the run (how far along you move)

* the vertical part of each step is called the rise (how far up you go).

When designing a staircase you need to take into account the legal requirements:

* the pitch (the steepness of the staircase) should not be more than 42° (the angle between the floor and the hypotenuse)

* the rise of each step cannot be more than 220mm.

The first measurement you need to take is the height of the total staircase, or total rise. Then you can work out the minimum number of risers allowed:

Total rise 2800mm ÷ 220mm (maximum height of risers) = minimum number of risers

If the total rise is 2800mm, and the pitch is 42° you can work out the total run using the tangent formula (TOA). You need to transpose this to make the length of the adjacent side the subject.

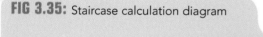

FIG 3.35: Staircase calculation diagram

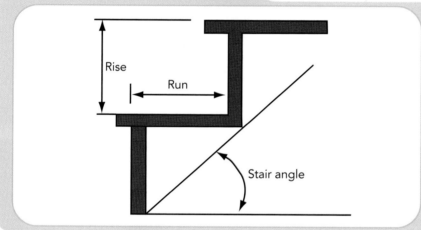

When you know the total run, you can work out the length of each step:

Total run ÷ number of risers = length of each step

You can also use Pythagoras' theorem to make sure that each step is square. The distance from the edge of one step to the edge of the next is the hypotenuse of a right-angled triangle.

$(\text{length of run})^2 + (\text{height of riser})^2 = (\text{hypotenuse})^2$

TRY THIS

1. If the pitch is of a staircase is 42° and the total rise is 2800mm, calculate the total run.

2. Calculate how much timber will be required to manufacture a staircase to the following measurements, where the step width is 600mm.
   a. Total rise 3.2 metres at an angle of 30°
   b. Total rise 4.7 metres at an angle of 27°.

Plan and design a pitched roof

A pitched roof is really two right-angled triangles back-to-back. The measurements you need are:

* the span – this is total width of the roof

* the total run – this is the base of one right-angled triangle, or half the span

* the total rise – this is the height of the roof

* the pitch angle – the angle between the total run and the hypotenuse.

FIG 3.36: Pitched roof

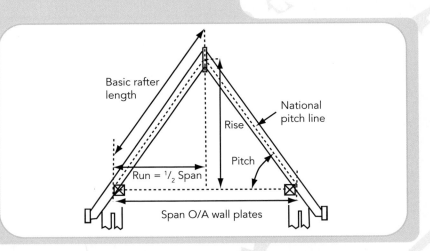

The pitch of the roof should be given in the building specifications. For this example the pitch is 25°. The first measurement you need to make is the span. This should be measured from the outside edge of one external wall to the outside edge of the opposite external wall. In this example, the span is 10m. Now you can work out the total run, half the span = 5m.

The pitch angle is adjacent to the total run, so you can work out the height of the roof using the tangent (TOA) formula.

Some buildings are bespoke and are not off-the-shelf designs. In order to calculate the amount of timber and roofing materials required we have to use all of the measurements. Figure 3.37 shows a simplified diagram of a roof and Figure 3.38 shows the measurements for this roof.

FIG 3.37: Simplified roof design

FIG 3.38: Roof measurements

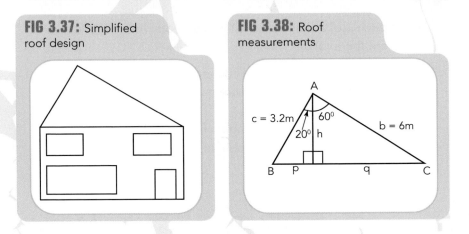

To calculate the area of the roof shown in Figure 3.38, we need to know its rise (h) and its span (p + q). We can use the sine (SOH) formula to work out p and q.

$$\sin 20° = \frac{p}{5}$$

$$p = 3.2 \times \sin 20° = 1.09m$$

$$\sin 60° = \frac{q}{6}$$

$$q = 6 \times \sin 60° = 5.20m$$

We can use the cosine (CAH) formula to work out the rise.

$$\cos 60° = \frac{h}{6}$$

$$h = 6 \times \cos 60° = 3m$$

We can now calculate the area of the roof section by using the formula for the area of a triangle.

$$\text{Area} = \frac{1}{2}(p + q) \times h = \frac{1}{2}(1.09 + 5.20) \times 3$$

$$\text{Area} = 9.4m^2$$

TRY THIS

Solve the following problem.

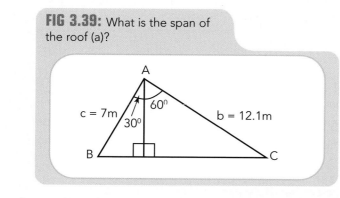

FIG 3.39: What is the span of the roof (a)?

Now that we have found all of the dimensions you can easily calculate the area of the roof in order to price up the materials needed.

Solving other construction problems using cosine, sine and tangent

To design and manufacture one-off roof trusses such as the one shown in Figure 3.40, you have to put together precise calculations

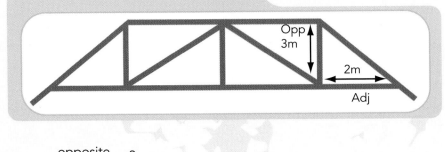

FIG 3.40: Roof truss

$$Tan\theta = \frac{opposite}{adjacent} = \frac{3}{2} = 1.5$$

Now find the angle that has a Tan of 1.5

Inv Tan = $Tan\theta^{-1}$ of 1.5 = 56.31°

TRY THIS

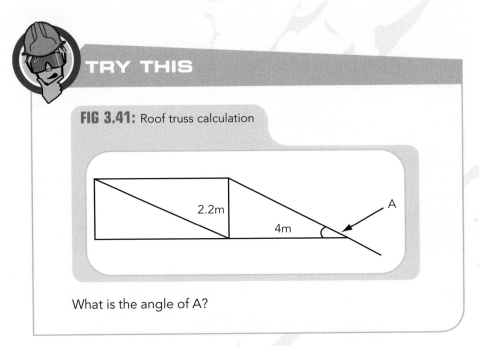

FIG 3.41: Roof truss calculation

What is the angle of A?

Setting out

The dimensions from a working drawing for a building are set out on the ground using pegs, markers and string. You need to make sure that you get the measurements right and check that the angles of corners are square.

Setting out buildings

A block plan shows:

* where a building will be positioned on site

* the shape and size of the building.

Measurements are marked on the plan to show:

* distances from the boundaries or other fixed points

* measurements of the building itself.

These measurements are used when setting out the perimeter wall.

To mark out the site wooden pegs are driven into the ground at corners. Nails are fixed on the tops of the pegs and a builder's line pulled taut from nail to nail to show the position of the wall.

FIG 3.42: Temporary peg

5

3

4

Temporary peg for marking 3:4:5 ratio

The tops of the pegs must be level for accurate measurements to be obtained. This is achieved by making the first peg (usually on the front line of the building) a datum peg. It is from this peg that all the other pegs are levelled. Levelling is done by putting a straight edge and spirit level across two pegs and hammering the second peg down until the bubble in the spirit level shows that the pegs are level.

Straight edge

Distances between corners are often more than the length of the straight edge and so temporary, intermediate pegs are used to transfer the level across to the next corner. Reverse the spirit level to check if it is accurate.

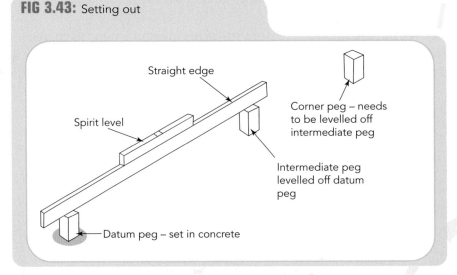

Straight edge

Spirit level

Corner peg – needs to be levelled off intermediate peg

Intermediate peg levelled off datum peg

Datum peg – set in concrete

Pegs are often positioned beyond corners so that the line crosses at the exact corner.

FIG 3.44: Pegs positioned beyond corners

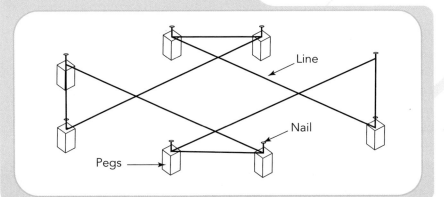

Line

Nail

Pegs

Pegging out

Most buildings are based on squares or rectangles. This means that corners will be right angles (90°). To make sure that lines are at right angles to each other, corners can be set out using the 3:4:5 method. This will always give a perfect right angle of measurement accurately.

FIG 3.45: Setting out a rectangle

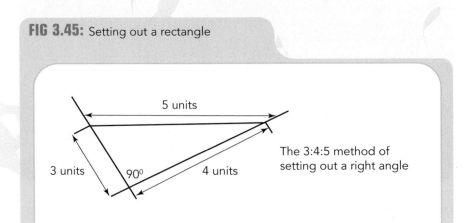

5 units

3 units

90°

4 units

The 3:4:5 method of setting out a right angle

TRY THIS

Find an unused open space and set out the perimeter of a rectangular building 20m x 15m.

WATCH IT!

Always check your calculations – they must be accurate – if not you could have a disaster on your hands.

Strategies for accuracy:

Approximate the answer to make sure you're on track, for example:

✱ 9.8 x 303 should be around 10 x 300 = 3000

✱ 79 ÷ 11.3 should be close to 77 ÷ 11 = 7

Make sure your decimal point is in the right place.

Do the calculation at least twice – if it's a difficult one, get someone else to check it.

Checking for square

An easy way of checking if a square or rectangle has been set out correctly, with all corners at 90°, is to check the diagonals across corners. If they are exactly the same length the setting out is correct.

FIG 3.46: Checking for square

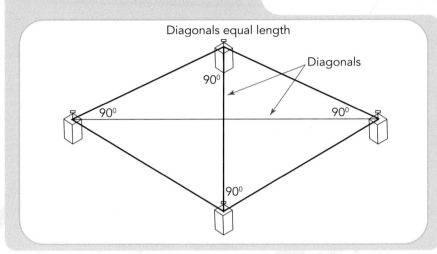

Diagonals equal length

Diagonals

90° 90° 90° 90°

Rounding off

Never round numbers down – you may not have enough materials if you do.

For example, if you need 2 x 4m, and 2 x 2.2m of skirting board, then the sum is 12.4m. If you rounded this down to 12m this would be 400mm too short.

The timber would probably be supplied in 4.8 metre lengths.

You need to order three lengths.

TRY THIS

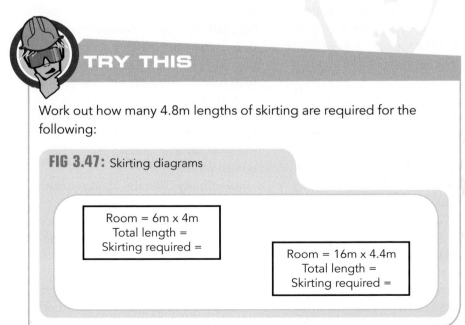

Work out how many 4.8m lengths of skirting are required for the following:

FIG 3.47: Skirting diagrams

Room = 6m x 4m
Total length =
Skirting required =

Room = 16m x 4.4m
Total length =
Skirting required =

⮎ The pitch of a staircase should not be more than 42°

⮎ The rise of each step cannot be more than 220mm

⮎ Most buildings are based on squares and rectangles

⮎ You can use simple trigonometry to solve length, width, area and angle problems

DIFFICULTY RATING   DATE

absorbent 101 – the ability of a material to take in and hold another material or substance.

accident 50 – an unplanned event that results in injury or ill health of people, or damage or loss to property, plant materials or the environment, or loss of a business opportunity.

accredited qualification 37 – a qualification recognised by the industry and issued by an approved organisation. For example, BTECs, A-levels and NVQs.

acetylene 73 – a highly flammable gas used in welding.

acute angle 125 – an angle of less than 90°.

aesthetic 11, 41 – pleasing to the eye, artistic.

anodising 93 – the electrical coating of aluminium with a protective layer of oxide.

apathy 65 – a don't care attitude.

apprenticeship 33 – a scheme to learn a trade or skill by a combination of paid work and part-time study.

biodiversity 42 – the range of different types of life form, plant and animal, which co-exist and interact in an area.

biomass 45 –the total amount of living material (plants and animals) in an area.

block release 37 – the release of students from work for blocks of time, such as a week or a term, to attend college.

breathing apparatus 74 – apparatus used to supply clean air to workers in confined spaces, or in dust-, smoke- or fume-filled areas.

brittleness 89 – the ease with which a material snaps or breaks.

brownfield site 29, 42 – an urban site that has been built on before and therefore has the potential for further development.

budget 41 – the amount of money allocated to pay for a particular job.

built environment 8 – anything constructed by humans to support human activity. For example, houses, schools, hospitals, airports, parks and open spaces, roads and bridges.

Cable Avoidance Tool (CAT) 78 – a device that can pick up the presence of metal. Often used with a generator to locate and map underground network of services.

carbon monoxide 77 – a colourless, odourless gas that can kill if inhaled in large quantities.

Cartesian coordinates 126, 129 – the x and y values for a point on a graph.

cash flow 15 – the money going into and out of a business.

chain of events 50 – all the things that happened in the lead-up to an accident and which were definitely connected with that accident.

change of state 96 – a physical change where the chemical structure of molecules stays the same but a substance may change from solid to liquid, liquid to gas and back again. For example, ice to water to water vapour and back to water.

close call 50 – an unplanned event that does note result in injury, illness or damage – but very easily could have. Also called a **near miss** or dangerous occurrence.

coefficient of expansion 101 – the change in length, area or volume of a material per unit change in temperature.

combustibility 89 – the ability to burn.

combustible 61, 73 – capable of burning, burns easily.

commercial 19 – relating to business – mainly shops, warehouses, showrooms and offices

common denominator 113 – a number in fractions that divides into all other **denominators**.

community 11 – all the people living in an area.

conduction 95, 97 – transfer of heart energy from one source to another.

consortium 15 – an association of business partners (could also include local or national government and banks).

Construction Skills Registration Scheme (CSCS) 35 – a scheme that provides a card listing the relevant training and competencies a person has to perform their job.

constant 107 – a number with a never changing value that is used to help in certain calculations. For example, **pi**.

contaminants 61 – anything that causes pollution.

contingencies 113 – anything that stops work going according to plan. For example, absence of a worker through illness.

contraction 97 – decrease in size.

convection 95, 97 – movement of heat in air; warmed air moves upwards.

Corgi registration 35 – a register of gas installation businesses employing qualified and competent tradespeople.

cuboid 107 – a six-sided, three-dimensional figure where each face is a rectangle.

day release 37 – the release of students from work for a day at a time to attend college

degradation 93 – the decay or disintegration of a material.

delegate 29 – pass the responsibility for a task or job to another person or organisation.

denominator 107 – the bottom number in a fraction.

density 84, 85 – the amount of mass per unit of volume of a material

dimension 107 – a category of measurement. For example, length, width or height.

economy 15 – the material and financial resources of a country.

environmental 53 – to do with the world around us and our impact on it.

environmental impact assessment (EIA) 11 – survey of the natural environment and how a construction project will affect it.

equilateral triangle 125 – one in which all the sides are the same length and all the angles are exactly 60°.

evaluation 113 – putting a value on something.

evaporation 95, 97, 101 – the **change of state** of a liquid into a vapour.

executives 23 – people with senior managerial responsibility in a business.

extrapolation 129 – finding values beyond the scope of the graph using the equation of the graph.

facilitate 11 – make easy, assist.

factor of safety (FOS) 87 – an amount factored into an equation to ensure the safety and stability of a structure when withstanding **loads**.

first aid 66 – the basic assistance that you can give without medical training to the victim of an accident.

flammable 73 – easily set on fire.

flora and fauna 45 – the plants and animals, including birds, insects and fish, that live in an area.

force 85 – anything which makes an object move. For example, **gravity**, **pressure** and **tension**.

gravity 85, 89 – the attractive **force** that one object exerts on another.

Gross Domestic Product (GDP) 13 – the activity of the national **economy** measured by the amount of goods and services produced, and the amount of income generated and money spent.

HASAWA 1974 57 – Health and Safety At Work etc Act 1974 sets out safety rules for both employers and employees.

hazards 58, 61 – dangers.

hazardous waste 70 – any waste materials on a building site that could be dangerous if not disposed of correctly. For example, asbestos, fluorescent tubes and paint containing lead.

HSE 57 – Health and Safety Executive.

health surveillance 41 – a programme of regular health checks for people working with hazards.

heat stress 64 – the adverse effect on humans of working in very high temperatures.

heritage 41 – of historic value, to be preserved.

hierarchy 61 – top-down structure from the most effective to the least effective.

hierarchy of controls 58 – a system of control selection that start with the most effective control, and if this is not possible moves on the next level of control and so on.

Hooke's law of elasticity 87 – the **strain** on an elastic material is related to the **stress** placed on it.

humidity 64 – a measure of the moisture content in the air.

hypotenuse 122, 125 – the longest side of a **right-angled triangle**, opposite the right angle.

induction 69 – initial company training for new workers.

Industrial Revolution 15 – a period following the invention of the steam engine in the late eighteenth century when machinery was introduced into industry.

infrastructure 11 – network of services such as roads, power lines, water pipes.

integration 19 – combining different parts into a working whole.

interpolation 127, 129 – interpreting values between points on a graph.

isolation – removing the source of electrical, gas, water or fuels supply so that equipment, machinery, pipework or cabling is completely inactive.

isosceles triangle 125 – one in which two angles and two sides are the same size.

joule 97 – a unit of energy.

lag 101 – to insulate an area or item, such a pipe or a tank.

latent heat (enthalpy) 96 – the that is absorbed or released when a substances **changes state**.

load 85, 89 – the overall force a structure is subjected to, including supporting weight and **mass** and outside forces, such as pressure from the wind.

mass 84 – the amount of matter (solid particles) in a compound.

mastics 61 – tile adhesives.

modulus of elasticity 87 – the amount of elasticity of an object or substance.

moment 89 – the turning effect produced by a **force** acting on a distant object.

near miss 50 – an unplanned event that does not result in injury, illness or damage – but very easily could have. Also called a **close call** or dangerous occurrence.

Newton (N) 89 – a unit of **force** named after Sir Isaac Newton.

obtuse angle 120, 125 – an angle that is greater than 90°.

parallel 107 – term used to describe two lines that are equal distance apart.

physical change 101 – a **change of state** that does not involve a chemical reaction.

pi (π) 107 – a **constant** used in calculations concerning circles.

pitched roof 132, 137 – a roof with an angle of more than 15°.

pivot 89 – the point about which an object rotates.

porosity 89 – the amount of spaces within a material that allow it to absorb liquid.

porous 101 – containing spaces through which liquid or air may pass.

potable 23 – describes water that is fit for humans to drink.

pressure 85 – continuous physical **force** exerted on or against an object.

private limited company (ltd) 21 – a company, such as a small business, that does not sell shares to the public.

professional development 33 – progressing your career by increasing your knowledge through courses, workshops and other forms of training.

prohibition notice 57 – a legal order to stop work.

protractor 125 – an instrument used for measuring angles.

public limited company (plc) 21 – a company that sells shares to the public on the **Stock Exchange**.

radiation 95, 97 – heat waves moving from a hot material into the surrounding air.

refurbishment 23 – renewing the internal fittings, fixtures and finishes of a building.

rendering 129 – applying a coat of mortar, usually to the outside of a building.

renovation 19 – restoring an old building to good condition, but not necessarily its original condition. Renovation often involves an updating of materials, appearance or use.

replenished 41 – re-stocked.

restoration 19 – bringing an old building back to its original state. For instance by repairing the brick, stone, timber, plaster and other work.

right-angled triangle 125 – one in which one angle is exactly 90°.

risk 58, 61 – the chance of danger.

risk assessment 57 – a survey of a job undertaken before anything is done to identify any undue risks to workers or the general public, plus any recommendations to minimise those risks.

scalene triangle 121, 125 – one in which every angle is less than 90°.

seismic 19 – movement of the earth caused naturally, for instance by volcanic action, earthquake tremors or subsidence, or by human action, for instance by explosions.

sensible heat 96 – the heat that you can feel and measure. The heat absorbed by a material which does not cause a **change in state**.

shoring 77 – support for a structure or excavation wall.

spall 101 – crumbling of the weathered face of a brick or block.

spatial 19 – relating to the space occupied by an object or the space available, and including – dimensions, shape and location.

stakeholders 19 – everyone who has an interest in a project.

Stock Exchange 23 – organisation based in the City of London where shares in businesses are traded.

strain 87 – the way in which **stress** changes the size or shape of a material. For example, when a spring is pulled it gets longer.

stress 87 – the force (per unit area) exerted on the internal structure of a material.

sustainablity 8 – making sure that whatever we do to the environment now does not damage it beyond repair in the future.

tempering 93 – the heat treatment of metal to increase hardness or strength.

tender 29 – an offer to contract a job at a specified cost or rate; a bid.

tension 85 – a stretching force.

thermal 97 – related to heat.

thermal movement 97, 101 – the change in original shape or size of a material due to temperature change.

thermostatic strip 101 – a strip made of two metals with different **coefficients of expansion**.

transposing 107 – rearranging, swapping things around.

unseasoned 101 – new timber that hasn't dried out properly.

urban 15 – relating to towns and cities, rather than the countryside.

urban regeneration 23 – renewing areas of towns and cities that have become rundown and unattractive.

variable 137 – an unknown quantity that can change, such as the number of litres of paint required to paint a wall.

viability 11 – ability to live and thrive

volume 84 – the amount of space that material occupies.

Index